Principles and Standards for School Mathematics

Navigating
through
Problem Solving
and
Reasoning
in
Grades 6–8

Susan N. Friel
Fran Arbaugh
Edward S. Mooney
David K. Pugalee
Tad Watanabe
Margaret S. Smith

Susan N. Friel
Grades 3–6 Editor

Peggy A. House
Navigations Series Editor

NATIONAL COUNCIL OF
TEACHERS OF MATHEMATICS

Copyright © 2009 by
The National Council of Teachers of Mathematics, Inc.
1906 Association Drive, Reston, VA 20191-1502
(703) 620-9840; (800) 235-7566; www.nctm.org

Library of Congress Cataloging-in-Publication Data

Navigating through problem solving and reasoning in grades 6-8 / Susan N. Friel ... [et al.].
    p. cm.—(Principles and standards for school mathematics navigations series)
  Includes bibliographical references.
  ISBN 978-0-87353-608-0
  1. Mathematics—Study and teaching (Middle school)—Activity programs. 2. Sixth grade
(Education)—Activity programs. 3. Seventh grade (Education)—Activity programs. 4. Eighth
grade (Education)—Activity programs. 5. Problem solving in children. 6. Reasoning in
children. I. Friel, Susan N.
  QA135.6.N3743 2008
  372.7--dc22   2008041611

The National Council of Teachers of Mathematics is a public voice of mathematics education,
providing vision, leadership, and professional development to support teachers in ensuring
equitable mathematics learning of the highest quality for all students.

Printed in the United States of America

TABLE OF CONTENTS

CONTENTS OF THE CD-ROM

Introduction

Table of Standards and Expectations, Process Standards, Pre-K–Grade 12

Supplemental Activities

TV Watching
Making the Data

Applets

Shape Cutter
Triangle Explorer
Plop It!
Squaring the Triangle

Blackline Masters and Templates

All the blacklines listed above, plus the following:
Centimeter Grid Paper
Quarter-Inch Grid Paper
Whose Tour Was Most Successful?
  Graph 1
  Graph 2
  Graph 3
  Graph 4

Readings from Publications of the National Council of Teachers of Mathematics

"Promoting Algebraic Reasoning Using Students' Thinking"
  Joyce W. Bishop, Albert D. Otto, and Cheryl A. Lubinski
  Mathematics Teaching in the Middle School

About This Book

"Problem solving is central to inquiry and application and should be interwoven throughout the mathematics curriculum to provide a context for learning and applying mathematical ideas." (NCTM 2000, p. 256)

Reasoning is fundamental to what it means to know and be able to do mathematics. Making and investigating mathematical conjectures, developing and evaluating mathematical arguments, and selecting and using various types of reasoning are the essence of the creative act of *doing* mathematics.

> The beauty and efficacy of mathematics both derive from a common factor that distinguishes mathematics from the mere accretion of information, or application of practical skills and feats of memory. This distinguishing feature of mathematics is called mathematical reasoning, reasoning that makes use of the structural organization by which the parts of mathematics are connected to each other, and not just to the real world objects of our experience, as when we employ mathematics to calculate some practical result. (Raimi 2002)

Every student, at every level of understanding, should have opportunities to engage in mathematical reasoning, and thus an emphasis on reasoning should pervade all instruction in mathematics.

The Organization of the Book

Developing mathematical reasoning is an essential goal of middle school mathematics. The introduction to this book provides an overview of mathematical reasoning and proof in middle-grades mathematics. Chapters 1–5 present investigations that are intended to strengthen students' reasoning and problem-solving skills in all the major content areas of the grades 6–8 mathematics curriculum. Each chapter focuses on reasoning in a different area: number, measurement, geometry, data analysis, and algebra. Activities in each chapter are designed to develop students' skill in using ideas from particular content areas to reason about problems in various contexts. Chapter 6 presents an exploration of the Pythagorean theorem to illustrate the possibilities for instruction that supports the early stages in the development of students' understanding of proof.

The structure of each investigation reflects a cyclical learning model with four stages:

- *Engage* introduces the investigation to the students and stimulates their interest in it, sometimes by guiding them in preliminary work.
- *Explore* launches the students in the investigation. Usually in pairs or small groups, they consider possible approaches and strategies, in the process asking questions, making decisions, experimenting, recording results, and proposing solutions. The teacher observes and facilitates this work by posing questions or making suggestions as the students describe, clarify, or provide evidence for their processes and their conclusions. The students listen carefully to, and

Many of the activities in this book also support core topics identified for emphasis in NCTM's *Curriculum Focal Points for Prekindergarten through Grade 8 Mathematics: A Quest for Coherence* (NCTM 2006). *Curriculum Focal Points* specifies by grade level essential content and processes that *Principles and Standards for School Mathematics* (NCTM 2000) discusses in depth by grade band.

Blackline Master

CD-ROM

Principles and Standards

Three different icons appear in the book, as shown in the key. One signals the blackline masters and indicates their locations in the appendix, another points readers to supplementary materials on the CD-ROM that accompanies the book, and a third alerts readers to material quoted from *Principles and Standards for School Mathematics*.

"Students can learn about, and deepen their understanding of, mathematical concepts by working through carefully selected problems that allow applications of mathematics to other contexts."

(NCTM 2006, p. 256)

ask for clarifications of, one another's explanations. The teacher helps them give structure to the mathematics that they are using and connect it with previously learned concepts.

- ***Evaluate*** looks back at what the students have accomplished. Both the students and the teacher engage in evaluation, which, though listed as a stage in its own right, is an activity that should occur throughout an investigation. The students should continually assess their own learning while the teacher evaluates their developing knowledge or skill, their application of new concepts or processes, and any significant changes in their thinking.

- ***Extend*** offers the students opportunities to apply their newly acquired concepts or skills in other contexts. The cycle begins again as the students engage with a new problem, explore it, and explain and evaluate their work on it.

Using the book

This book includes an accompanying CD-ROM that provides readings for teachers' professional development, supplemental investigations for students, and applets for learning experiences that promote students' skill in reasoning. Teachers can allow students to use the applets in conjunction with particular activities or apart from them, to extend and deepen students' understanding. The resources on the CD support the fundamental idea that "students can learn about, and deepen their understanding of, mathematical concepts by working through carefully selected problems that allow applications of mathematics to other contexts" (NCTM 2000, p. 256). An icon in the margin (see the key) alerts readers to all materials on the CD.

Activity sheets for students appear as reproducible blackline masters in the appendix of the book, along with solutions to the problems. A second icon signals all blackline pages, which teachers can also print from the accompanying CD.

Throughout the book, margin notes supply teaching tips and pertinent ideas from *Principles and Standards for School Mathematics*. A third icon alerts the reader to these quotations, which highlight the importance of problem-solving skills to an understanding of the power and utility of mathematics.

The authors of this book do not attempt to provide a complete curriculum for problem solving and reasoning in the middle grades. The activities presented here are intended to be illustrative rather than comprehensive, and students should encounter them in conjunction with other instructional materials in relevant contexts. Clearly, reasoning and problem solving are integral parts of any mathematics program and cannot be addressed in isolation—on "problem-solving Fridays," for example. The purpose of this book is merely to highlight the nature of reasoning and problem solving, offering you and your students opportunities to see connections across content areas in mathematics.

Navigations Series

Grades 6–8

Problem Solving *and* Reasoning

Introduction

Every student has the potential to engage in mathematical reasoning, and an emphasis on reasoning should pervade all mathematical activity. Middle-grades students engage in two types of mathematical reasoning: inductive reasoning and deductive reasoning. Students have been reasoning in both ways from the early grades, but now they are ready to examine and categorize the two kinds of reasoning and use them more purposefully.

Inductive Reasoning

Inductive reasoning involves looking for patterns and making generalizations. Students in grades 6–8 would be likely to use this type of reasoning in considering the following problem (adapted from Smith et al. 2004):

> Ms. Green's class is planning to raise rabbits for the spring science fair. The students have 24 feet of fencing with which to build a rectangular pen to enclose the rabbits. The fencing comes in 1-foot segments that cannot be cut into smaller sizes. The students want the rabbits' pen to have as much room (area) as possible.
>
> *a*. How long should the students make each of the sides of the pen? What will the area of the pen be?
>
> *b*. Suppose that the students have 36 feet of fencing. How long should they make each of the sides of the pen? What will the area of the pen be?

"People untaught in mathematical reasoning are not being saved from something difficult; they are, rather, being deprived of something easy." (Raimi 2002)

c. Suppose that the students have 30 feet off fencing. How long should they make each of the sides of the pen? What will the area of the pen be?

d. How could Ms. Green's students go about determining the pen with the greatest area for any amount of fencing? How could they organize their work so that anyone who looks at it can understand it?

The students in your class might begin solving this problem by drawing pictures or constructing rabbit pens with square tiles. They would be likely to use familiar approaches and make guesses, perhaps computing the areas of different possible pens.

Fig. **0.1**.

Two pens using 24 feet of fencing

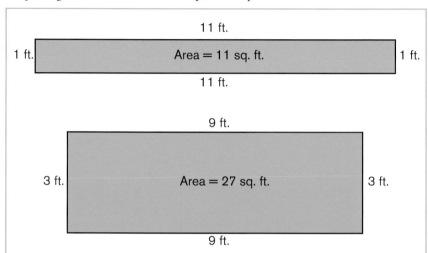

Eventually, the students might organize their information in a table like that in figure 0.2. Displaying their data in this way can help them see how the rectangular areas grow to 36 square feet. The table can also assist them in noting that the rectangles appear to "repeat" after this point. They might observe that as the rectangles grow smaller, the measurements for length and width interchange from earlier rectangles. With the visual aid of the table, the students are likely to discover that the rectangle with the largest area is a square.

However, will it always be the case that the largest rectangle with a given perimeter is a square? Students who continue to work on parts

Fig. **0.2**.

A table displaying dimensions of rectangles with perimeters of 24 feet

Perimeter (ft)	Width (ft)	Length (ft)	Area (ft2)
24	1	11	11
24	2	10	20
24	3	9	27
24	4	8	32
24	5	7	35
24	6	6	36
24	7	5	35
24	8	4	32
24	9	3	27

(*b*), (*c*), and (*d*) of the problem might reason inductively that of all the rectangles that they could make with a given perimeter, the square, or the most "square-like" rectangle, is the one with the greatest area.

Encountering problems such as this one can help middle school students appreciate the usefulness of inductive reasoning. They discover that inductive reasoning enables problem solvers to discern patterns, make educated guesses, formulate conjectures, test them, and generalize results.

Eventually, however, students in grades 6–8 also need to recognize the limitations of inductive reasoning. As *Principles and Standards for School Mathematics* (NCTM 2000) asserts, "Because many elementary and middle-grades tasks rely on inductive reasoning, teachers should be aware that students might develop an incorrect expectation that patterns always generalize in ways that would be expected on the basis of the regularities found in the first few terms" (NCTM 2000, p. 265). Inductive reasoning helps problem solvers develop an understanding of a problem but may be insufficient to lead them to the correct solution or give an explanation of a solution that is in fact correct, putting it beyond dispute. Generalizations obtained through inductive reasoning often must be proven through deductive reasoning.

Deductive Reasoning

In contrast to inductive reasoning, which depends on observing patterns and making conjectures on the basis of them, deductive reasoning depends on logic. Problem solvers make a logical argument, draw logical conclusions, and apply generalizations based on logic to specific situations. When middle school students have determined that for a fixed perimeter the rectangle with the largest area is a square, their "proof" may consist of an explanation of the inductive process that they used. One way to argue deductively that a square yields the largest rabbit pen is to make a graph of the equation

$$Area = length \times width,$$

where

$$Length = \frac{perimeter}{2} - width,$$

and, by substitution,

$$Area = \left(\frac{perimeter}{2} - width \right) \times width.$$

In the case of a rectangular pen with a fixed perimeter consisting of 24 feet of fencing, this means that

$$Area = (12 - width) \times width, \ \text{or} \ Area = 12(width) - (width)^2.$$

Figure 0.3 shows a graph of this function. The graph is a parabola with roots of 0 and 12. These points are not themselves part of the graph for the problem; no rectangular pen can have a width of 0 feet or 12 feet. At these widths, the rectangle collapses on itself; in other words, neither of these widths allows for the creation of a rectangle with a perimeter of 24 feet.

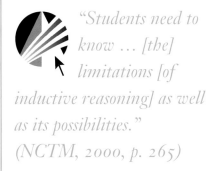

"Students need to know … [the] limitations [of inductive reasoning] as well as its possibilities."
(NCTM, 2000, p. 265)

Middle school students reason both inductively and deductively. Although they use argument and justification to make their points, they do not formalize either inductive or deductive proof. Nevertheless, they do make deductive arguments that could be considered "proofs."

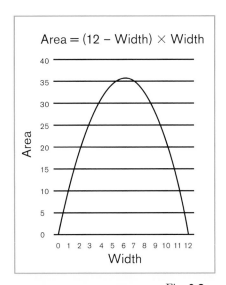

Fig. **0.3.**

A graph of the equation
$Area = 12(width) - (width)^2$,
for $0 < width < 12$

"Worthwhile mathematical tasks are those that do not separate mathematical thinking from mathematical concepts or skills, that capture students' curiosity, and that invite students to speculate and to pursue their hunches."

(NCTM 2007, p. 33)

The peak, or *vertex*, of the parabola always occurs exactly halfway between the two roots. In this case, the peak occurs when the width of the pen is equal to 6 feet. This provides the maximum area possible under the condition of a fixed perimeter of 24 feet.

By using inductive and deductive reasoning, students can judge for themselves the accuracy of their solutions to many different problems. As they learn mathematical concepts and solve problems, they become aware of the power of reasoning to unlock ideas in mathematics and phenomena in the everyday world.

Using Mathematical Tasks to Promote Mathematical Reasoning

A fundamental responsibility of all teachers is to select and present worthwhile tasks and materials, with a clear goal of creating opportunities for students to learn and reason about mathematics. *Mathematics Teaching Today* (NCTM 2007), an update of *Professional Standards for Teaching Mathematics* (NCTM 1991), recommends that teachers "design learning experiences and pose tasks that are based on sound and significant mathematics" (p. 32), characterizing worthwhile mathematical tasks as ones that—

- engage students' intellect;
- develop mathematical understandings and skills;
- stimulate students to make connections and develop a coherent framework for mathematical ideas;
- call for problem formulation, problem solving, and mathematical reasoning;
- promote communication about mathematics;
- represent mathematics as an ongoing human activity;
- display sensitivity to, and draw on, students' diverse background experiences and dispositions. (pp. 32–33)

In selecting worthwhile tasks, teachers consider factors from three areas: the mathematics content, the students' knowledge of the content, and the ways in which the students learn the content. The development of reasoning and problem-solving skills is an important goal of most mathematical tasks that teachers present to students.

Distinguishing among mathematical tasks is important. Not all tasks provide the same opportunities for developing students' thinking and learning. Tasks that call on students to perform a memorized procedure in a routine manner offer opportunities at one level. Tasks that engage students actively in thinking about concepts and making connections among them afford different opportunities at higher levels. The tasks that students encounter constitute their experiences of *what* mathematics is and *how* people "do" it (Doyle, 1983; 1988). Over the years, the cumulative effect of classroom-based tasks is the formation of the students' ideas about the nature of mathematics, whether mathematics is something that they personally can make sense of, and how long and how hard they are likely to need to work to do so.

The publication of *Curriculum and Evaluation Standards for School Mathematics* (NCTM 1989) and *Principles and Standards for School Mathematics* (NCTM 2000) has given rise to many attempts to describe the types of tasks that promote and develop students' understanding of mathematical concepts. Although the language used to describe worthwhile activities differs, the discussions share a common focus on tasks that have the potential to develop the students' capacity for non-algorithmic thinking.

Stein and colleagues (2000) provide a taxonomy of mathematical tasks based on the kinds and level of thinking required to perform them. Tasks place *high-level* cognitive demands on students by engaging them in the processes of active inquiry and validation or by encouraging them to use procedures in ways that they can connect meaningfully to concepts or understanding. "Active inquiry and validation" involve students in "doing mathematics," and *procedures* include standard and nonstandard processes, formulas, and algorithms. By contrast, tasks that place *lower-level* cognitive demands on students are those that encourage students to use mathematical procedures in ways that disassociate them from meaning or rely entirely on memorization or the reproduction of previously mastered facts.

The 1999 Video Study completed by the Third International Mathematics and Science Study (TIMSS) makes a similar distinction between low-level tasks, which students perform by *using procedures* (including basic computations) and high-level tasks, which they perform by *making connections* (focusing on concepts and discovering links among them) (NCES 2003; Stiegler and Hiebert 2004). Figure 0.4 provides examples of both types of tasks. Tasks that make high-level cognitive demands on students have a greater potential to develop their ability to reason mathematically and understand what constitutes mathematical proof.

Stein, Grover, and Henningsen (1996) provide a framework for tracking the cognitive demands of mathematical tasks as they unfold in instruction and for exploring the connection between instruction and learning. Their framework distinguishes three phases through which mathematical tasks pass in the instructional process. In phase 1, teachers consider the tasks in curricular or instructional materials (for example, on the printed pages of textbooks, ancillary materials, or materials created by teachers themselves). In phase 2, teachers set up the tasks or announce them in the classroom. Finally, in phase 3, students and teachers actually implement the tasks. All of these phases, but especially the implementation phase, have an important impact on what students learn.

This research indicates that the cognitive demands of a task can change between phases of the framework. For example, a task that appears in curricular or instructional materials is not always identical to the task that a teacher sets up in the classroom, and this task, in turn, is not always identical to the task that the students actually do. Between the setup and implementation phases, tasks can change for a variety of reasons, many of which have to do with the press of the classroom environment. It is interesting to note, however, that when teachers set up a task at a low level, students nearly always implement it as intended (Stein, Grover, and Henningsen, 1996).

"A mathematical investigation is a multidimensional exploration of a meaningful topic, the goal of which is to discover new ways of thinking about the mathematics inherent in the situation rather than to discover particular answers. Mathematical investigations demand that students speculate, conjecture, and generalize." (Chapin 1998, p. 333)

"Good tasks are ones that do not separate mathematical thinking from mathematical concepts or skills, that capture students' curiosity, and that invite them to speculate and to pursue their hunches." (NCTM 1991, p. 25)

 Henningsen (2000; available on the CD-ROM) shows that students whose teachers frequently set mathematical tasks at high cognitive levels learn more than their counterparts whose teachers frequently assign tasks at low cognitive levels.

 Smith and Stein (1998) and Stein and Smith (1998) provide more detail about the classification and design of mathematical tasks.

 Rachlin (1998) describes principles of task design that can be applied to increase the cognitive demand of a task from a low to a high level.

Low-Level Cognitive Demand

Task: Take a timed test on multiplication facts

Materials: None

Complete the following multiplication facts in one minute or less:

$2 \times 3 =$ ___ $5 \times 6 =$ ___ $10 \times 4 =$ ___

$4 \times 8 =$ ___ $8 \times 10 =$ ___ $4 \times 7 =$ ___

$9 \times 3 =$ ___ $3 \times 5 =$ ___ $6 \times 6 =$ ___

$6 \times 7 =$ ___ $7 \times 8 =$ ___ $2 \times 7 =$ ___

$3 \times 9 =$ ___ $8 \times 7 =$ ___ $9 \times 4 =$ ___

Task: Identify the factors of given products

Materials: None

a. What factor paired with 9 gives a product of 63?

b. What factor paired with 14 gives a product of 140?

Task: Find common factors and GCFs of given numbers

Materials: None

In problems 1–4 below, list the common factors for each pair of numbers. Then find the greatest common factor for each pair.

1. 18 and 36   2. 9 and 25

3. 60 and 45   4. 49 and 14

High-Level Cognitive Demand

Task: Set up a rectangular space with a given area

Materials: Tiles

Every year, the Jamestown Arts and Crafts Fair has an area for exhibits. The exhibitors use carpet squares to mark out their spaces. Each carpet square measures 1 square foot, and every exhibitor's space must have a rectangular shape. Rocks and Things Jewelry wants to rent a space that is 24 square feet.

a. Using 24 square tiles to represent the carpet squares, find all the possible ways in which the owner of Rocks and Things Jewelry can arrange the squares to make a rectangle. Use a table to record the length and width of each rectangle. Begin with a rectangle with a width of 1 foot.

b. Which of the options would you recommend, and why?

c. How are the rectangular possibilities that you found and the factors of 24 related? Explain your reasoning.

Task: Explore the product of two numbers and their GCFs and LCMs

Materials: None

1. Choose two whole numbers, A and B. Find the GCF and the LCM of each number. Do this for several pairs of numbers. Here are some examples:

A	B	GCF	LCM
8	28	4	56
7	23	1	161

What is the relationship between the product $A \times B$ and the GCF and LCM of A and B?

2. Look at the prime factorization for each of two whole numbers A and B. How are the prime factorizations related to the numbers' GCFs and LCMs? Connect this relationship to your findings in problem 1?

Fig. **0.4**

Examples of low- and high-level cognitive tasks

The Nature of Mathematical Reasoning and Proof

Proof is a vital component of mathematics and should be part of all students' mathematical experiences at all levels. The students' work in early grades emphasizes proof as an explanation of the structure of a pattern. Proof activities, especially those based on patterns, grow in value as students progress through the elementary, middle, and high school years. Waring (2000) distinguishes six levels in the development of students' understanding of proof, beginning with their appreciation of the need for proof, followed by their growing understanding of the nature of proof, and culminating in their acquisition of skills for constructing proofs. Figure 0.5 shows the articulation of Waring's six-level framework for this development.

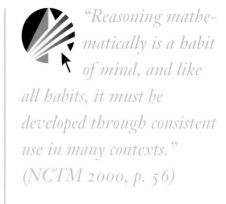

Proof Level 0: Pupils are ignorant of the necessity for, or even existence of, proof.

Proof Level 1: Pupils are aware of the notion of proof but consider that checking a few special cases is sufficient as proof.

Proof Level 2: Pupils are aware that checking a few cases is not tantamount to proof but are satisfied that either

    a) checking for more varied and/or randomly selected examples is proof; or

    b) using a generic example forms a proof for a class.

Proof Level 3: Pupils are aware of the need for a generalised proof and, although unable to construct a valid proof unaided, are likely to be able to understand the creation of a proof at an appropriate level of difficulty.

Proof Level 4: Pupils are aware of the need for, and can understand the creation of, generalised proofs and are also able to construct such proofs in a limited number of, probably familiar, contexts.

Proof Level 5: Pupils are aware of the need for a generalised proof, can understand the creation of some formal proofs, and are able to construct proofs in a variety of contexts, including some unfamiliar.

Fig. **0.5.**

Waring (2000, p. 11) differentiates six levels in the development of students' understanding of proof.

At proof level 0, students have no appreciation of the notion of proof. However, as they move up from this level and progress through five more levels, they gradually develop an understanding of the generalized nature of proof and its role in justifying conjectures. They also gradually become capable of constructing proofs in a variety of contexts.

At proof level 1, students recognize the existence of proof but have no appreciation of its generalized nature. Proof level 2 marks an important transition. Some students now replace a few particular cases checked at proof level 1 with a greater number of particular cases that are either more varied or randomly selected; other students use a generic example to represent a class of cases.

Students who appreciate the generalized nature of proof have attained proof level 3. They can follow a short chain of deductive reasoning, but they are not necessarily able to construct formal proofs. However, they can distinguish between proof and practical demonstration.

At proof level 4, students can construct proofs in a limited range of contexts, including ones that are familiar or give rise to "intuitive" thought processes or reasoning that a student can express informally. Students at proof level 5 have developed a deeper understanding of the nature and role of proof and can construct proofs in a variety of contexts, possibly using some formal language where appropriate.

The purpose of focusing on informal and, eventually, formal mathematical proof is to give students opportunities to learn that mathematical statements can—and, whenever possible, should—be proved or explained. Students gradually develop an appreciation of the need for justification or proof through experiences of such reasoning in the classroom. In the process, many will realize that checking a statement for one special case is insufficient as proof. They will discover that they should either treat a special case as a generic example of a general statement or consider additional cases that are varied or randomly selected.

Some students may be capable of understanding the use of simple algebraic and geometric arguments as part of generalized proofs. Whatever the students' level of understanding of the details, repeated exposure to proofs or explanations and justifications helps them recognize that new mathematical knowledge can be derived from familiar facts and properties.

Instruction in the middle grades should emphasize the development of understanding at levels 1 and 2, with possible extensions to level 3. The focus should be on fostering an awareness of the necessity for proof, and tasks should more often call for justification or explanation than for "proof" in a more formal sense. The Reasoning and Proof Standard outlined in *Principles and Standards* calls for students at all levels to have many and varied experiences that allow them to reason as they—

- examine patterns and structures to detect regularities;
- formulate generalizations and conjectures about observed regularities;
- evaluate conjectures;
- construct and evaluate mathematical arguments. (NCTM 2000, p. 262)

Students in grades 6–8 should develop arguments to support their conclusions in solving problems involving a variety of mathematical topics. Although mathematical argument in the middle grades usually lacks the formality and rigor customarily associated with mathematical proof, it should include the formulation of plausible conjectures, testing of those conjectures, and presentation of associated reasoning for evaluation by others.

"Students should discuss their reasoning on a regular basis with the teacher and with one another, explaining the basis for their conjectures and the rationale for their mathematical assertions." (NCTM 2000, p. 262)

PROBLEM SOLVING *and* REASONING

Chapter 1
Reasoning about Number and Operations

"One important decision that each teacher must make is how best to engage students in thinking and reasoning about mathematics. Indeed, what is the interplay among knowledge, skills, and mathematical reasoning?"
(NCTM 1999, p. vii)

Students' knowledge of number and number systems grows over many years of learning. Ideally, students develop this knowledge gradually, by reasoning in the process of completing many varied mathematical tasks that make high-level cognitive demands on them. The reasoning that students do in such contexts often involves identifying the problem to solve, formulating an appropriate strategy to use in solving it, representing and analyzing information, and monitoring and evaluating possible solutions.

Students become engaged in this process explicitly when they are working in a context that motivates them to "do mathematics"—for example, when they are solving the problem presented in the introduction about the largest rectangular rabbit pen that someone can make with 24 feet of fencing. Students who routinely make sense of mathematics apply this process regularly and automatically in their day-to-day learning about number.

Effective instruction helps students make sense of concepts and procedures as they develop understanding. They build meaning from what they know. For instance, in teaching your students to add fractions, you might encourage them to use benchmarks to check the reasonableness of their answers. Figure 1.1 illustrates the reasoning process that students might follow in this case.

Fig. **1.1.**

Making sense of the addition of two fractions by comparing each to $\frac{1}{2}$ as a benchmark. (Reprinted from *Principles and Standards for School Mathematics* [NCTM 2000, p. 219].)

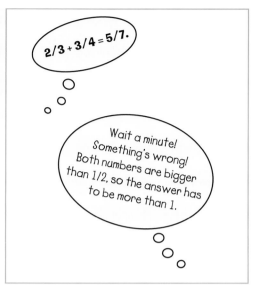

Similarly, you should offer your students numerous opportunities to use reasoning to understand relationships among fractions, decimals, and percentages. Figure 1.2 shows a task that affords one such opportunity, calling on students to justify the equivalence of 75 percent and $\frac{3}{4}$.

Fig. **1.2.**

A task in which students use reasoning to show that 75 percent is equivalent to $\frac{3}{4}$. (Reprinted from *Principles and Standards for School Mathematics* [NCTM 2000, p. 215].)

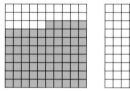

Use the drawing above to justify in as many different ways as you can that 75% of the square is equal to 3/4 of the square. You may reposition the shaded squares if you wish.

To support your students' understanding of number in the middle grades, you should engage them in tasks that develop their ability to use their conceptual knowledge fluently, efficiently, accurately, and flexibly. For example, you might write on the board

$$12 \times 5 \times 7, \ 3 \times 5 \times 21, \ \text{and} \ 3 \times 25 \times 7.$$

Then you might say, "You know that $3 \times 5 \times 7$ is equal to 105. How can you use this fact to find these other products?" Problem solving that requires middle-grades students to investigate, conjecture, and communicate their findings supports their development of *number sense*.

Number sense is a general understanding of number and operations, together with an ability to use this understanding in flexible ways to make mathematical judgments and develop useful strategies for solving complex problems (Burton 1993; Reys 1991). Simple, informal investigations of numbers and their relationships promote middle-school students' development of number sense while preparing them for later, more sophisticated explorations in number theory. Middle-grades students can reason about numbers and number concepts in a multitude of tasks, including searching for patterns in sums of numbers and examining multiples, factors, and prime numbers. The two investigations in this chapter, Reasoning about Sums of Consecutive Numbers and Reasoning about Factors and Multiples, offer such opportunities.

"Many teachers fear that if they increase the level of challenge in assignments, they will leave students behind. Our premise is that moderate challenge actually promotes students' engagement." (Turner, Styers, and Daggs 1997 [available on the CD-ROM], p. 67)

Reasoning about Sums of Consecutive Numbers

Goals

- Work with concepts related to consecutive addends, divisibility, and powers of 2

- Make conjectures, determine their validity, and support or contradict them by making mathematical arguments

- Explain problem-solving strategies and justify solutions

Materials and Equipment

For each student—

- A copy of each of the following activity sheets—

- "Sum-thing about Consecutive Numbers"

- "Recording Sheet for 'Sum-thing about Consecutive Numbers'"

For each team of two or three students—

- A sheet of chart paper

- Two or three marking pens

For the teacher—

- (Optional) An overhead projector and one or two transparencies

pp. 126–29, 130–31

Prior Knowledge

Students should be able to add, multiply, and divide integers.

Learning Environment

For the most part, the students work in pairs or groups of three, although they sometimes begin independently. The teacher serves as a facilitator, asking questions to support the students' mathematical reasoning about patterns that they observe. The students formulate, write, and evaluate conjectures that they believe will always be true.

Discussion

In this activity, students investigate the counting numbers from 1 to 35 to determine which are sums of consecutive counting numbers. In the process, they find patterns and make conjectures about odd and even numbers and powers of 2.

To introduce the activity, orchestrate a brief discussion of the meaning of *consecutive numbers*. Discuss the fact that the term usually refers to two or more integers in a sequence in which the *absolute difference* between any two "neighboring" terms is 1. Be sure that the students understand that the absolute difference between two numbers is the distance between the numbers on a number line without respect to the

This activity is adapted from Driscoll (1999).

direction traveled from one number to the other. The students should recognize that absolute differences are always positive quantities.

Also emphasize that in the activity the students will be working with *counting numbers*. They should know that the set of counting numbers begins with 1, contains the numbers customarily used for counting in the everyday world, and is written in set notation as {1, 2, 3, …}.

Engage

On the board or an overhead transparency, write the three sums that appear at the beginning of the activity sheet "Sum-thing about Consecutive Numbers":

$$3 + 4 = 7$$
$$2 + 3 + 4 = 9$$
$$4 + 5 + 6 + 7 = 22$$

Ask your students to supply other examples of sums of consecutive numbers, restricting themselves to the counting numbers. Record several responses. Give each student a copy of the activity sheet.

Explore

Have your students begin by working independently on step 1. They must find all the ways in which they can express the counting numbers from 1 to 35 as the sums of two or more consecutive counting numbers. After they have worked on their own for 5 to 10 minutes, assign each student one or two partners and direct the teams to continue working on step 1. Allow the students enough time to complete the task.

Step 2 asks the students if they discovered any patterns in the consecutive addends that they found for sums from 1 to 35. When all the teams are ready for this step, distribute copies of the blackline master "Recording Sheet for 'Sum-thing about Consecutive Numbers.'" Direct the students to use the sheet to organize their work from step 1. Now they must organize the strings of consecutive addends that they discovered in step 1 according to the number of addends in a string. Figure 1.3 shows a slightly different recording sheet completed by a team of students in one classroom for the numbers 1–35.

Note that these students did not initially restrict the consecutive addends to the counting numbers. Thus, they represented 1 as the sum of consecutive whole numbers 0 and 1. They expanded to the integers in finding four consecutive addends that give a sum of 2: –1, 0, 1, 2. However, after using 0 one more time in a string of four addends that yield a sum of 6, the students confined the remainder of their work to the counting numbers. The addends given for the sum of 2 (–1 + 0 + 1 + 2 = 2) will be of interest if you later ask your students to extend their work to the integers.

Also give each team a sheet of chart paper and explain that they will use it to display their discoveries in a way that will allow them to share their findings with the class. Detail any specific format that you want them to use for this process. If some of your students are stymied at this point, suggest that they reorganize their data—for example, making lists of numbers that are the sums of two consecutive addends, three consecutive addends, and so on.

Evaluate

Gather the students together as a group to share their discoveries. Make a visible record of all their observations and conclusions by

Number	Two Numbers	Three Numbers	Four Numbers	Five Numbers	Six Numbers	Seven Numbers
1	0+1=1					
2			-1+0+1+2=2			
3	1+2	0+1+2				
4						
5	2+3					
6		1+2+3	0+1+2+3			
7	3+4					
8						
9	4+5	2+3+4				
10			1+2+3+4			
11	5+6					
12		3+4+5				
13	6+7					
14			2+3+4+5			
15	7+8	4+5+6		1+2+3+4+5		
16						
17	8+9					
18		5+6+7	3+4+5+6			
19	9+10					
20				2+3+4+5+6		
21	10+11	6+7+8			1+2+3+4+5+6	
22			4+5+6+7			
23	11+12					
24		7+8+9				
25	12+13			3+4+5+6+7		
26			5+6+7+8			
27	13+14	8+9+10			2+3+4+5+6+7	
28						1+2+3+4+5+6+7
29	14+15					
30		9+10+11	6+7+8+9	4+5+6+7+8		
31	15+16					
32						
33	16+17	10+11+12			3+4+5+6+7+8	
34			7+8+9+10			
35	17+18			5+6+7+8+9		2+3+4+5+6+7+8

Fig. **1.3.**

Students organize sums from 1 to 35 according to the numbers of consecutive addends in the sums.

posting their chart-paper displays or using a section of the board to list all their discoveries. Ask probing questions as the students present their work to ensure that they are reasoning clearly and can explain their thinking in a way that makes sense to themselves and others. Question the listeners as well as the presenters to be certain that they understand what the presenters are saying.

Encourage your students to discuss any patterns that they see in their recording sheets. The comments of some students follow:

- "The first pattern we noticed was that when two numbers were added to find a sum, every other row [in the chart] worked; when three numbers were added, every third number [row] worked, and so on. The second pattern we found was that once you know what cell has a solution, you can then take away the first number in the sequence and add the next at the end of the sequence [for example, for 10 you have 1 + 2 + 3 + 4, and for the fourth number after ten, 14, you have 2 + 3 + 4 + 5]."

- "Only the odd numbers can have two consecutive numbers. We [also] found that once you know what cell has a solution, you can take away the first number in the sequence and add the next to the end of the sequence."

- "Any number divisible by 3 has a solution of 3 consecutive addends, so every third number can be made with three addends."

A possible way to have your students evaluate their own and their classmates' discoveries is to assign each small group one discovery. Ask the students to construct a mathematical argument that supports or contradicts the conjecture.

This part of the activity can play an important role in developing students' abilities to reason mathematically. Challenging students to see beyond offering a few examples to validate an argument takes them in the direction of formal reasoning and proof. Encourage your students to generalize in making their arguments, and when they have finished, bring them together to share their work and collaborate on a list of valid conjectures.

Extend

You can use steps 3 and 4 on the activity sheet "Sum-thing about Consecutive Numbers" to give your students an opportunity to apply their validated conjectures to counting numbers beyond those used in the original exploration.

It is important to consider the numbers that cannot be sums of consecutive numbers. These numbers are the powers of 2: {1, 2, 4, 8, 16, 32, …}. Take the opportunity to write these numbers in exponential form: $2^0 = 1$, $2^1 = 2$, $2^2 = 4$, $2^3 = 8$, $2^4 = 16$, $2^5 = 32$, and so on.

Furthermore, some students in your class may observe that some of the numbers—multiples of 4—can always be written as sums of two consecutive odd numbers: 4 = 1 + 3, 8 = 3 + 5, 12 = 5 + 7, 16 = 7 + 9, 20 = 9 + 11, and so on. Eighth-grade algebra students may be able to investigate this fact as a challenging extension of the activity (see fig. 1.4). They could expand this exploration to all integers.

Fig. **1.4**.

Showing that any multiple of 4 can
be expressed as the sum of two consecutive
odd integers

A multiple of 4 can be written as $4 \times m$, or $4m$, where m is any integer.

All even integers are divisible by 2, so an even integer can be written as $2n$, where n is any integer. Odd integers are not divisible by 2, so an odd integer can be written as $2n + 1$.

The absolute difference between two consecutive odd integers is 2, so the sum of two consecutive odd integers can be written as $(2n + 1) + (2n + 3)$.

If it is true that any multiple of 4, or $4m$, can be the sum of two consecutive odd integers, then

$(2n + 1) + (2n + 3) = 4m$, for some integers n and m.

$(2n + 1) + (2n + 3) = 4n + 4 = 4(n + 1)$

Because m can be written as the sum of some integer n plus 1, $4(n + 1)$ is a multiple of 4.

In discussing shortcuts for determining whether a number is the sum of two or more consecutive addends, students may make an important observation: an odd number that is the product of two odd numbers other than 1 and itself may be the sum of one or both of those numbers of consecutive addends. In other words, the factors of an odd number, other than 1 and itself, may point to numbers of consecutive addends for which the number can be the sum. For example, 45 is equal to the product of 5 and 9 and thus can be written as the sum of five 9s (or nine 5s):

$$45 = 9 + 9 + 9 + 9 + 9$$
$$45 = 5 + 5 + 5 + 5 + 5 + 5 + 5 + 5 + 5$$

Working from the "middle" addend and adjusting those to the left by subtracting first 1, then 2, then 3, and so on, as necessary, and those to the right by adding 1, then 2, then 3, and so on, gives two strings of consecutive addends:

$$45 = 7 + 8 + 9 + 10 + 11$$
$$45 = 1 + 2 + 3 + 4 + 5 + 6 + 7 + 8 + 9.$$

Because of the restriction to the counting numbers that the problem imposes, the method is only partially successful in the case of 21. The number 21 can be written as the sum of three 7s or seven 3s:

$$21 = 7 + 7 + 7 = 6 + 7 + 8$$
(string of three consecutive numbers)

$$21 = 3 + 3 + 3 + 3 + 3 + 3 + 3 = 0 + 1 + 2 + 3 + 4 + 5 + 6$$
(string of seven consecutive whole numbers but only six consecutive counting numbers with the elimination of 0).

In fact, a useful extension of the original activity could begin with another question: "What patterns do you discover if you use integers instead of counting numbers as your consecutive addends for sums from 1 to 35?" In the next activity, the students move from considering addends and sums to exploring factors and multiples.

Reasoning about Factors and Multiples

Goals

- Examine, differentiate and apply the concepts of least common multiple (LCM) and greatest common factor (GCF)
- Apply the concept of the LCM in finding lowest common denominators and the concept of the GCF in simplifying fractions
- Understand the relationships among two numbers, their multiples and factors (including prime factors), and their LCM and GCF

Materials and Equipment

For each student—

- A copy of each of the following activity sheets:
  - "Looking for the Least"
  - "In-Venn-stigating Multiples and Factors"
  - "What Are the Relationships?"

For selected pairs or groups of three students—

- A sheet of chart paper or one or two transparencies
- Several marking pens

For the teacher—

- Two packages of small sticky notes (different colors, if possible)
- One or two marking pens
- (Optional) An overhead projector and one or two transparencies
- (Optional) Several sheets of chart paper

pp. 132–33, 134–41, 142–43

Prior Knowledge

Students should have some experience with multiples and factors and be acquainted with the concepts LCM, GCF, and prime factorization.

Learning Environment

The students work on three sets of problems, in pairs or groups of three for the most part, although they sometimes begin their work independently.

Discussion

The three parts of this activity take the students through related sets of tasks. In part 1 ("Looking for the Least"), they apply their knowledge of multiples and least common multiples as they explore story problems set in everyday contexts. In part 2 ("In-Venn-stigating Factors and Multiples"), they use Venn diagrams as representational tools to sort multiples and factors of pairs of selected numbers. In part 3 ("What Are the Relationships?"), they supply several examples that they explore to make a conjecture about the relationship of paired numbers to their GCF and LCM. In examining this relationship, they discover other useful ones as well.

Part 1—"Looking for the Least"

The activity sheet "Looking for the Least" presents story problems in three contexts. The first setting is an upcoming school field trip that will include a hot dog lunch. Given the different numbers of hot dogs and rolls in standard packages, the students investigate the smallest numbers of packages of hot dogs and rolls that will be necessary for the trip. The second setting is the seashore, where lights on two lighthouses rotate at different rates. The students determine the shortest time that elapses between moments when the lights are synchronized in their arcs. In the third setting, the students examine research findings on different amounts of sleep needed each night by a child, a teenager, and an adult. The students discover the smallest number of nights that each must sleep for all to have slept the same total number of hours.

Engage

To introduce the story problems on the activity sheet, present the following sample problem, which describes a simplified situation for riders on two Ferris wheels:

> Riverside Amusement Park has two Ferris wheels. Melissa is going to ride on the big wheel, and Mandy is going to ride the small wheel at the same time. They know that the big wheel makes one complete revolution in 9 minutes, and the small wheel makes one revolution in 6 minutes. If Melissa and Mandy begin their rides at exactly the same time, what is the shortest time that the rides must last for both girls to arrive at the bottom again at the same time?

You can display the problem on the board or an overhead projector if you wish. Let students talk about the problem and think about strategies for solving it. One way to visualize the situation is to show the times on a line, as illustrated in the margin.

In 18 minutes, the big wheel makes 2 revolutions, and the small wheel makes 3 revolutions. Both riders are now in their original positions—in tandem again for the first time since boarding their respective Ferris wheels.

When your students have reached this point in their reasoning, ask them to consider how the numbers 6, 9, and 18 are related. Guide them in understanding that 18 is a *common multiple*—in this case, the *least common multiple*, or LCM—of the numbers 6 and 9. If your students are not well acquainted with the term *least common multiple*, emphasize that *least* means *lowest* or *smallest*. Some middle school students may initially think that *least* modifies *common* and may interpret the phrase *least common* to mean "most unlikely" or "rarest" as a modifier for *multiple*. Stress that *least*—by itself—modifies the phrase *common multiple*. The least common multiple—the LCM—is simply the smallest or lowest common multiple. (Other common multiples include 36, 54, 72, etc.)

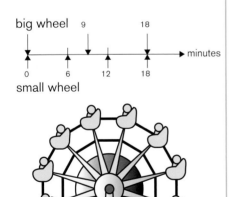

Navigating through Problem Solving and Reasoning in Grades 6–8

Explore

Give each student a copy of the activity sheet "Looking for the Least," and explain that the students will be solving story problems that call for the kind of thinking that they used in solving the problem about the Ferris wheels. Assign each student one or two partners, and encourage the teams to find more than one way to think about a problem. Invite them to try to model each problem mathematically so that they can use mathematics to arrive at a solution. Observe them closely as they work, identifying two or three teams to present their solutions to the class. Give these students transparencies or chart paper on which to display their work.

Evaluate

When everyone has finished the problems on the activity sheet, call the class together, and let your selected teams present their solutions. After discussing these in depth for each problem, ask if any students found other ways to solve the problem. Observe both the presenters and the listeners carefully to be sure that they are reasoning with understanding.

Extend

Your students have now had an opportunity to solve story problems that involve two or more rates and call for using the LCM to determine the *least* or *fewest* in a particular context. To extend their skill, you can bring them back to the context of Ferris wheels, this time giving them a more complex problem. Introduce them to several world-famous Ferris wheels with different periods of revolution. For example, the Yokohama (Japan) Ferris wheel completes a revolution in 15 minutes, the Everland Ferris wheel (Seoul, South Korea) completes a revolution in 10 minutes, and the Navy Pier Ferris wheel (Chicago, Illinois) completes a revolution in 7.5 minutes. Say to the students, "Suppose that these three Ferris wheels are in the same amusement park and that three people board them at exactly the same time. How long will the rides need to last before all three riders arrive together again at the bottom of the wheels?"

To solve this problem, students can experiment with a diagram like that for the simpler problem relating to Ferris wheels in the "Engage" section. A sample diagram for this more complicated problem appears in margin.

Some students might solve the new problem by reasoning in a slightly different way. The Yokohama Ferris wheel revolves once for every two revolutions of the Navy Pier Ferris wheel. So these two wheels will need to revolve for 15 minutes before riders who board them at the same time will arrive at the bottom at the same time. But the Everland Ferris wheel revolves three times for every two revolutions of the Yokohama Ferris wheel. So these two Ferris wheels will need to revolve for a total of 30 minutes before riders who boarded them at the same time will arrive at the bottom at the same time. Because the Navy Pier Ferris wheel completes two revolutions for every one revolution of the Yokohama Ferris wheel, it will complete four revolutions in 30 minutes while the Yokohama Ferris wheel completes two revolutions and the Everland Ferris wheel completes three revolutions. Thus, rides on all three wheels will need to last 30 minutes before riders who leave the ground at the same time will arrive back on the ground at the same time.

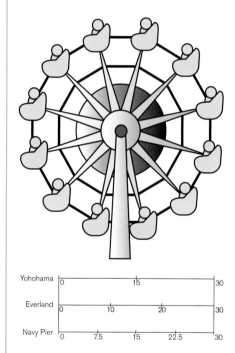

Yohohama	0		15		30
Everland	0	10		20	30
Navy Pier	0	7.5	15	22.5	30

To offer your students a different extension of their work with rates and multiples, you might use the following question posed by a student to the online "Ask Dr. Math" Web page at the Math Forum:

> My dad was helping me with my math and ended up confusing me more than I was to begin with with this problem:
>
>> The Plush Hotel received a shipment of glasses packed in full cartons of 40 glasses each. Another shipment of glasses packed 24 to a carton went to the Maison Restaurant. Glasses were also shipped to State University, but this shipment contained cartons with 25 glasses each. If the hotel, the restaurant and the university each received the same number of glasses and if none of them received more than a thousand glasses, how many glasses were in each shipment? How many cartons were in each shipment?
>
> My dad came up with 600 glasses: hotel = 15 cartons; restaurant = 25 cartons; and university = 24 cartons. But I have no idea how he got there and now he has left for work. Please help. (http://mathforum.org/library/drmath/view/58498.html)

Ask your students to take the role of Dr. Math and write an explanation that will help this baffled student understand how to solve the problem. Tell your students to be sure to explain why the father's solution is correct and how he may have reasoned to arrive at it. To see the explanation that "Ask Dr. Math" actually offered online, go to the Web address given at the end of the extract.

Part 2—"In-Venn-stigating Multiples and Factors"
"In-Venn-stigating Multiples and Factors" introduces the students to the use of two-circle Venn diagrams as visual tools for sorting and analyzing multiples and factors of pairs of numbers. They examine their completed diagrams to identify the LCM or GCF of the numbers. They reason about the relationships between the paired numbers to characterize sets of common multiples.

Engage
Choose two numbers—say, 6 and 8. Write 10 multiples of each on small sticky notes. (If you have sticky notes in two colors, use one color for multiples of 6 and the other for multiples of 8.) Draw a large two-circle Venn diagram on chart paper or the board, labeling one circle "Multiples of 6" and the other "Multiples of 8." Distribute your sticky-note multiples to twenty students. Ask these students to come forward, one at a time, and place each sticky note in the correct location on the Venn diagram. If a student is not sure where to place his or her multiple, solicit help from the class.

When two students have sticky notes for a common multiple, have them place the two notes one on top of the other in the intersection of the circles in the Venn diagram, indicating that the two multiples "count as one" because they are "common" to both numbers. (If the two notes are different colors, overlap them in the intersection so that both colors show, indicating that the number is a multiple of both 6 and 8.)

Continue until all the multiples have been placed correctly. Then discuss the information that the Venn diagram provides about the multiples of each number, the common multiples of the two numbers, and their least common multiple.

Explore

Give each student a copy of the activity sheet "In-Venn-stigating Multiples and Factors," and assign each student one or two partners. Let the teams work by themselves to complete steps 1–9 on the sheet. In these steps, the students complete Venn diagrams by sorting multiples of numbers that—

- have no common factors other than 1 (3 and 8; 2 and 7);
- are either factors or multiples of each other (5 and 15; 4 and 8);
- have at least one factor in common in addition to 1 (8 and 12; 14 and 21).

At several points in the activity (steps 3, 6, and 9), the students reflect on their work and make generalizations about the multiples in the intersections on the basis of the relationships between the numbers whose multiples they are diagramming.

As you did while your students were completing the story problems in part 1, move around the room and observe them closely. Select several teams to display their work on chart paper or overhead transparencies to share with the class. Ask one team to prepare Venn diagrams to show their work in steps 1 and 2, a second team to prepare diagrams for steps 4 and 5, and a third team to show their work in steps 7 and 8.

Evaluate

When everyone has completed steps 1–9 on the activity sheet, invite your selected teams to come forward, display their work, and report on it. Use the students' answers to the questions in steps 3, 6, and 9 to summarize characteristics of the numbers in the given pairs and their multiples as shown in the students' completed Venn diagrams. As in part 1, observe both the presenters and the listeners carefully to be sure that they are reasoning with understanding.

Extend

Your students' discussion of their work up to this point should have touched on *factors* as well as multiples of the numbers in each pair. To extend these ideas and make them explicit, have your students work again in their teams to complete steps 10–13—the final steps on the activity sheet. Steps 10 and 11 direct the students to make Venn diagrams showing the factors of 8 and 12 and the factors of 14 and 21. In each case, the students must tell how the information displayed in the diagram helps them determine the greatest common factor (GCF). Then in steps 12 and 13 they make Venn diagrams showing the *prime factors* of the same pairs of numbers. They look back at their work, finding both the LCM and the GFC that they identified for each pair. Their work in part 2 culminates in the formulation of an explanation of how the information in their latest Venn diagrams—those showing prime factors—helps them find both the LCM and the GFC easily and directly. To cement your students' understanding of what they have learned, move on to part 3.

Part 3—"What Are the Relationships?"

The relationships of two numbers and their factors, prime factorizations, GCF, and LCM are useful for students to examine and understand. The activity sheet for part 3 provides examples of several pairs of numbers and their respective factors, GCFs, prime factorizations, and LCMs. The students offer additional examples and look for patterns in the relationships.

Engage

Write a 4 and a 6 on the board. These numbers constitute the first pair given as an example on the activity sheet "What Are the Relationships?" Ask the students for the LCM and the GCF of 4 and 6. When they have supplied 12 as the LCM and 2 as the GCF, write these numbers on the board. Tell the students that an interesting relationship exists between paired numbers and their GCF and LCM. You may have some students who quickly see the relationship—that the product of the selected numbers, 4 and 6, is equal to the product of their GCF and LCM. If students do notice this relationship, ask if they think it always holds, and if so, can they explain why? If no one notices the relationship, let it remain hidden as the students explore other examples in the activity to see if they can find it.

Next, ask the students to give you the factors of 4 and 6. Use set notation in writing the factors on board: {1, 2, 4} for 4 and {1, 2, 3, 6} for 6. Ask, "Does seeing the factors in this way help you identify the GCF?" Some students may say something like, "Both sets have a 1 and a 2, and the 2 is bigger, so it's the GCF." Ask the students to continue to think about the relationship between the GCF and the elements of the two sets as they work. Gradually, they should begin to think explicitly in terms of the *intersection* of the two sets, and observe that the greatest factor in the intersection is the GCF.

Ask the students to give you the prime factorizations of 4 and 6: 2×2 for 4, and 2×3 for 6. Ask, "Does taking a look at the prime factorizations help you identify the LCM?" Write the numbers in each prime factorization as the elements of a set—{2, 2} for 4 and {2, 3} for 6. Some students may notice, even if they cannot state the fact explicitly, that the LCM is the product of the elements in the union of the two sets. If your students appear to have no clear sense of a relationship, do not reveal it at this point. Ask the students to think as they work about the connection between these two sets and the LCM.

Explore

Give each student a copy of the activity sheet "What Are the Relationships?" and let the students work on it independently, in pairs, or in groups of three. The sheet gives five examples of pairs of numbers, including the pair discussed above (4 and 6). It shows their factors, GCFs, prime factorizations, and LCMs. After supplying and working through six new number pairs in addition to the examples on the sheet, the students may conjecture that the product of a pair of numbers is equal to the product of the numbers' LCM and GCF.

If your students make this conjecture, they should consider whether it is true in all cases. They should test other number pairs. With guidance, they can use Venn diagrams again, this time to show the factors in the prime factorizations of paired numbers. Figure 1.5 presents a sample Venn diagram for the paired numbers 8 and 28. Such representations can help persuade them that this relationship in fact always holds, although at this stage they may not understand *why*.

In the process of examining Venn diagrams of the factors in two numbers' prime factorizations, the students may notice two other useful relationships, suggested in the "Explain" section above:

- The LCM of two numbers is the product of the prime factors in the *union* of the sets shown in the Venn diagram.
- The GCF of two numbers is the product of the prime factors in the *intersection* of the sets unless the intersection is empty, in which case the GCF is 1.

The number 1, which is not a prime number, is the only factor that two numbers have in common in such a case. If the students notice these connections between the LCM and the union, on the one hand, and the GCF and the intersection, on the other, they can use them to evaluate their conjecture and understand why the product of the paired numbers is equal to the product of their GCF and LCM.

Number Pair

8 and 28

Prime Factorizations

$2 \times 2 \times 2 = 8$
$2 \times 2 \times 7 = 28$

Let set A contain the factors in the prime factorization of 8:
$A = \{2, 2, 2\}$.
Let set B contain the factors in the prime factorization of 28:
$B = \{2, 2, 7\}$.

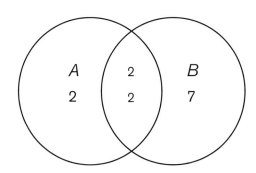

A and $B = \{2, 2\} = A \cap B$
GCF $= 4 =$ Product of the elements in $A \cap B$

A or $B = \{2, 2, 2, 7\} = A \cup B$
LCM $= 56 =$ Product of the elements in $A \cup B$

Product of the Paired Numbers $=$
$8 \times 28 = 2 \times 2 \times 2 \times 2 \times 2 \times 7 = 224$

Product of GCF $\times$ LCM $= 4 \times 56 = 2 \times 2 \times 2 \times 2 \times 2 \times 7 = 224$

$8 \times 28 = 4 \times 56$

Fig. **1.5**.

Relationships of the paired numbers 8 and 28 to their GCF, LCM, and the factors in their prime factorizations

Two students who were summarizing their ideas from their work in a whole-class discussion had the following exchange with their teacher:

First Student:	We tried 4 and 6. We found that 4 times 6 equals 24, right? And 2 is the GCM and 12 is the LCM, and 2 times 12 equals 24.
Second Student:	And so the GCF and the LCM times each other equals the pair of numbers times each other.
Teacher:	*[addressing the class]* Can every team try that with their number pair? Just because it works once doesn't mean it works for every number. *[after waiting for the students' results]* So we found twelve instances that their findings are true—so can we say this is always true?
First Student:	We haven't tried any prime numbers. What about 2 and 5?
Teacher:	Okay, the GCF is 1, and the LCM is 10. What is the product of the GCF and the LCM, and what is the product of the two numbers? 10? … Oh, it works. Do we have a prime and a nonprime? Give me an example.
Second Student:	Yes, 9 and 7 … and 6 and 5.
Teacher:	Well, what is the GCF? Hmmm, 1 again … I wonder if that matters. Can we think of a number where the GCF is not 1 and one of the numbers is prime?
First Student:	How about 5 and 10? That works. 5 is a factor of 10. That happens with 8 and 4, too. The GCF is the factor, and the LCM is the other number, so the products have to be the same.

When your students have tested enough cases to conjecture that the product of the GCF and the LCM is always equal to the product of two paired numbers, ask them to share their thoughts about why this might be so. You may find that your students go back and forth between testing cases, making a claim that the two products are the same, and then deciding to test another case.

Turn the discussion to what it would take to convince everyone that this is always true and no further testing is necessary. With your assistance, they may be able to generalize their work in a Venn diagram such as that shown in figure 1.6. Such a diagram can help show that the prime factorization of the product of the GCF and the LCM is the same as the prime factorization of the product of the two selected numbers, so the two products are equal.

Extend

Opportunities for integrating reasoning into other basic number contexts are readily available. For example, you might have your students look at divisibility "rules"—the informal tests of divisibility that they

Fig. **1.6.**

Using a Venn diagram to demonstrate that the product of two numbers X and Y is equal to the product of their GCF and LCM

Number Pair

X and Y

Prime Factorizations

Suppose that $a \times b \times c \times d \times e$ is the prime factorization of X and $d \times e \times f \times g \times h$ is the prime factorization of Y.

Let set A contain the factors in the prime factorization of X:
$$A = \{a, b, c, d, e\}.$$

Let set B contain the factors in the prime factorization of Y:
$$B = \{d, e, f, g, h\}.$$

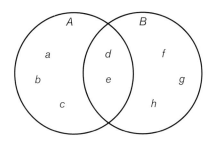

A and $B = \{d, e\} = A \cap B$
GCF $= d \times e =$ Product of the elements in $A \cap B$

A or $B = \{a, b, c, d, e, f, g, h\} = A \cup B$

LCM $= a \times b \times c \times d \times e \times f \times g \times h$
$=$ Product of the elements in $A \cup B$

Product of the Paired Numbers $= X \times Y$
$= (a \times b \times c \times d \times e) \times (d \times e \times f \times g \times h)$

Product of GCF $\times$ LCM
$= (d \times e) \times (a \times b \times c \times d \times e \times f \times g \times h)$

The students can rearrange the factors
to see that the products are equal.

have learned gradually and perhaps unconsciously in grades 3–5. Figure 1.7 shows a number of these rules. The students could start with the following: A number is divisible by 5 if the digit in the units place is 0 or 5. Challenge your students to explain why this rule "works."

Using basic number knowledge, the students can decompose any whole number into the sum of its place values. For example,

$$4265 \div 5 = (4000 + 200 + 60 + 5) \div 5$$
$$= (4000 \div 5) + (200 \div 5) + (60 \div 5) + (5 \div 5).$$

They can reason that any multiple of 10 is divisible by 5: the factors of 10 are 2 and 5, so any multiple of 10 has a factor of 5. Thus, they know that 4000, 200, and 60 are all divisible by 5. If, as in the example,

Fig. **1.7.**

Divisibility "rules" for divisors 2–10, except 7

Divisor	Characteristics of a number that is evenly divisible by the divisor
2	The number is even. (That is, the units digit is 0, 2, 4, 6, or 8.)
3	The sum of the digits is divisible by 3.
4	The last two digits form a number that is divisible by 4.
5	The units digit is 5 or zero.
6	The number is divisible by both 2 and 3.
8	The last three digits form a number that is divisible by 8.
9	The sum of all the digits is divisible by 9.
10	The units digit is zero.

the digit in the units place is a 5 instead of a 0, the number is divisible by 5 because 5 has itself as a factor and is divisible by itself.

If a whole number has a 0 or a 5 in the ones' place, then the students know that the number is a multiple of 5, has 5 as a factor, and is divisible by 5. The inverse is also true: if the digit in the units place is not a 0 or a 5, then the number is not a multiple of 5, does not have 5 as a factor, and is not divisible by 5. For additional extensions, ask your students to reason about and explain other rules in figure 1.7.

Conclusion

Proficiency in numerical reasoning includes the ability to analyze numerical relationships logically and discover underlying principles. Using such skills routinely not only makes middle school students better problem solvers but also brings them into contact with the forms and methods of proof. Exposure to proof as a natural part of problem solving is important. Referring to work by Collins and colleagues (1989), *Principles and Standards* asserts,

> Reasoning and proof are not special activities reserved for special times or special topics in the curriculum but should be a natural, ongoing part of classroom discussions, no matter what topic is being studied. In mathematically productive

Navigating through Problem Solving and Reasoning in Grades 6–8

classroom environments, students should expect to explain and justify their conclusions. When questions such as, What are you doing? or Why does that make sense? are the norm in a mathematics classroom, students are able to clarify their thinking, to learn new ways to look at and think about situations, and to develop standards for high-quality mathematical reasoning. (NCTM, 2000, p. 342)

The activities explored in this chapter have dealt with some aspects of number theory—a topic that affords many opportunities for reasoning, conjecturing, and proof. Proof in these contexts usually takes the form of informal argumentation and justification: students customarily generate examples and then reflect on them to develop generalizations. The investigation in the following chapter extends this work into the area of measurement, showing how students in the middle grades can use what they know about the area of a rectangle to reason about the areas of other shapes, including a parallelogram, a triangle, and a trapezoid.

PROBLEM SOLVING *and* REASONING

Chapter 2
Reasoning about Measurement

"In the middle grades, students should have frequent and diverse experiences with mathematical reasoning as they:

• *examine patterns and structures to detect regularities;*

• *formulate generalizations and conjectures about observed regularities."*

(NCTM 2000, p. 262; regularities is used informally to indicate common characteristics, resemblances, or relationships.)

What role does reasoning play in middle school students' learning of measurement? What patterns and relationships can students at this level become aware of in measurement activities? How might they use the processes of conjecture and generalization to build understanding in this important area of the curriculum?

Students develop their understanding of many measurement concepts in the middle grades. They investigate new attributes such as rate and other derived measurements. They learn about precision and error in measurement.

In addition, they discover for themselves many important formulas for the area and volume of various geometric shapes. Much of this work builds on ideas that the students encountered in grades 3–5. In the intermediate grades, they learned about the concept of area and discovered that the area of a rectangle is equal to the rectangle's length times its width. In the middle grades, their understanding of this relationship grows and becomes the basis for their investigations of the areas of other shapes.

The investigation in chapter 2 demonstrates the important role that reasoning can play in learning with understanding formulas for the areas of a parallelogram, a triangle, and a

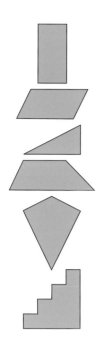

trapezoid. The investigation shows how each new formula builds on previous ideas, and it also guides students in extending and reinforcing their learning by exploring the area of a kite and a "stair"—presented as a closed figure that looks something like the outline of a staircase, with steps that become higher at a constant rate (see the lowest shape in the group in the margin). The chapter's investigation is likely to extend over several class periods and can be separated into parts to suit specific classroom needs.

"Whenever possible, students should develop formulas and procedures meaningfully through investigation rather than memorize them."
(NCTM 2000, p. 244)

Navigating through Measurement in Grades 6–8 (Bright et al. 2005) is a useful resource for a variety of measurement tasks that can engage students in reasoning and problem solving.

Reasoning about Area Formulas

Goals

- Use an understanding of the area of a rectangle to reason about the areas of particular parallelograms and apply techniques of decomposition, transformation, and recomposition to discover that $A = b \times h$ gives the area of any parallelogram
- Build on an understanding of the area of a parallelogram to reason about and derive the formula for the area of a triangle
- Work from an understanding of the areas of a parallelogram and a triangle to reason about and derive the formula for the area of a trapezoid
- Apply the formulas for the areas of a rectangle, a parallelogram, and triangle in determining the areas of other shapes

Materials and Equipment

For each student—

- A copy of each of the following activity sheets:
  - "Parallel(ogram) Universe"
  - "Moving into Triangular Territory"
  - "How Much Area Does a Trapezoid Trap?"
  - "Carryover to Kites"
  - "Stepping Up to Stairs"

For each small group of three or four students—

- Extra copies of the figures on the activity sheets for decomposing and recomposing shapes
- One or two pairs of scissors

For several groups of students—

- (Optional) Poster board or chart paper

Prior Knowledge

Students should have had opportunities in grades 3–5 to "develop, understand, and use [the] formula" for the area of a rectangle to find the areas of a variety of rectangles, as *Principles and Standards for School Mathematics* (NCTM 2000, p. 170) recommends.

Learning Environment

In each section of the investigation, the students work in groups of three or four to reason together and share ideas before all the groups reconvene as a class to compare their methods and discoveries.

The sequence of tasks in this investigation can be classified as *procedures with connections.*

pp. 144–45, 146–47, 148–50, 151–52, 153

Discussion

Students entering middle school typically bring a good understanding of the area of a rectangle and how to calculate it. Their work in grades 3–5 should have helped them develop a very concrete sense of the area of a rectangle as the length of the rectangle multiplied by its width. Often they have learned this fact through hands-on experimentation with square tiles or grids of squares or dots. Students usually begin grade 6 knowing that the area of a rectangle is $A = l \times w$, even if they do not express the idea in this way, in a formula with symbols.

In middle school, students often use this understanding to discover that the product of the base and the height of a parallelogram is its area (see fig. 2.1*a*). This discovery typically provides a starting point for explorations of the areas of other shapes, especially a triangle and a trapezoid.

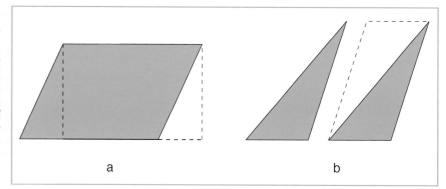

Fig. **2.1.**

Reasoning about the area of (*a*) a parallelogram by decomposing it and recomposing the parts as a rectangle of equal area and (*b*) a triangle by copying it and rotating the copy to compose a parallelogram with twice the area. (Reprinted from *Principles and Standards for School Mathematics* [NCTM 2000, p. 244].)

a   b

Each section of this investigation challenges the students to extend their reasoning as they progressively determine area formulas by understanding the area of a given shape in terms of the area of one or more familiar shapes whose areas they know how to find. After helping the students explore the area of a parallelogram on the basis of their previous work with rectangles, the investigation invites them to use their newly acquired understanding to reason about the area of a triangle (see fig. 2.1*b*). This work enables them to write an expression of the general formula, often given as $A = \frac{1}{2}(b \times h)$.

As the students work through the successive sections of the investigation, they identify patterns in their methods of understanding one shape in relation to another, and they use these to develop area formulas for more complicated shapes. The students apply their work with rectangles, parallelograms, and triangles to trapezoids, reasoning about the area of a trapezoid (see fig. 2.2) and deriving the general formula, usually given as $A = \frac{1}{2}(b_1 + b_2)h$. Figure 2.2 shows two ways in which students can work with an isosceles trapezoid to understand its area.

The investigation fixes the students' attention squarely on different ways of reshaping the areas of specific figures to reason about their areas in terms of the areas of more familiar shapes. These methods, techniques, and strategies—not the derivations of the formulas—are the focus of the work. In fact, the investigation assumes that many students will already have some familiarity with the formulas without understanding why they "work." If you are using the investigation as an introduction to the development of the formulas, consider giving your students

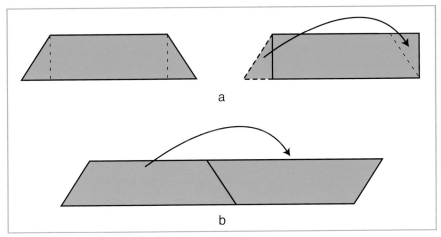

a

b

Fig. **2.2.**

To determine the area of isosceles trapezoid, students can (*a*) decompose it by cutting a right triangle from one side, sliding it, and flipping it to form a rectangle of equal area, or they can (*b*) copy the trapezoid and rotate the copy to form a parallelogram with twice the area. (Reprinted from *Principles and Standards for School Mathematics* [NCTM 2000, p. 244].)

The applet Shape Cutter, available on the CD-ROM, lets students use the methods of decomposition, transformation, and recomposition that they discover in the investigation.

additional sample figures, some on grids and others on plain paper with measurements specified, to provide more opportunities for experimentation and computation.

The accompanying CD-ROM includes the applet Shape Cutter, which allows students to connect dots on a grid to draw shapes (parallelograms, triangles, trapezoids, and kites are all possible, for example). Then the students can make cut lines through their shapes and slide, turn, or flip the pieces to make other shapes that preserve the areas of the original shapes. Your students can work with this applet at any stage of the investigation to reinforce or extend their understanding of the use of decomposition, transformation, and recomposition to explore the areas of shapes.

Engage

Divide your students into pairs or small groups and ask each team to think of as many mathematical and scientific formulas as they can and make a list of them. Then have the groups share their lists and record the formulas on the board (or an overhead transparency). Their suggestions may include the following:

- Distance = rate × time
- Area of a rectangle = length × width
- Area of a square = (side)^2
- Degrees Celsius = (degrees Fahrenheit – 32) × $\frac{5}{9}$

Question your students about the formulas:

- "How did mathematicians and scientists come up with these formulas?"
- "What makes the formulas valuable?"
- "Why do the formulas work?"

Explain that formulas are *generalizations* that grow out of *conjectures*, which in turn rest on *reflections on the steps that someone must take to determine a particular result*.

If your students have not listed the formula for the area of a rectangle, ask them about it, and add it to the list. Write the formula as "Area = length × width," or $A = l \times w$. Then ask, "Why does this formula work?" Listen to what your students say. Be sure that they communicate the idea that when they measure the length and width of a rectangle

Formulas are *generalizations* that grow out of *conjectures*, which in turn rest on *reflections on the steps that someone must take to determine a particular result*.

with the same linear unit—a centimeter, for example—they can use their measurements to determine how many unit squares fit along the length and the width of the rectangle. The measure of the width of the rectangle tells the number of unit squares that will fit along the side, and the measure of the length tells the number of columns of unit squares that will extend from top to bottom of the rectangle (see fig. 2.3). Multiplying these measurements yields the number of unit squares in the rectangle.

Fig. **2.3.**

Why does the formula $A = l \times w$ work?

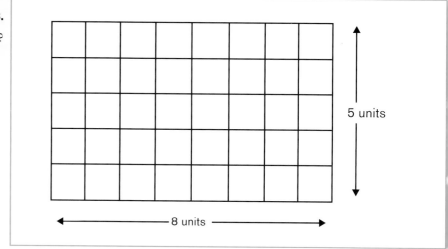

5 units

8 units

Give each student a copy of the activity sheet "Parallel(ogram) Universe." Ask, "Do you think that knowing how to find the area of a rectangle can help you find the area of a parallelogram?" Tell the students that the activity sheet will help them explore this question. Have them continue to work in pairs or in small groups, and remind them that they should be thinking together—not just checking one another's answers. Step 1 presents two parallelograms on grids (shown in the margin). Be sure to provide extra copies of these figures for your students to cut into pieces and rearrange.

As your students work, encourage them to focus on the actions that they carry out. Research indicates that learners construct mathematical understanding through their reflections on their actions—both mental and physical (see, for example, Kamii [1985] and Simon and colleagues [2004]).

Although in many instances students benefit from organizing their findings in a table, such an approach can actually short-circuit the fruitful process of reasoning and reflection that the investigation invites. Students who determine the areas of parallelograms and record their results in a table like that in figure 2.4 may readily conjecture that the area of any parallelogram is the product of the base and the height. However, they may have no idea *why* that product gives the area. Focusing only on the measurements that they obtain rather than on their actions may prevent them from understanding why the computational process works.

By contrast, students who actually reconfigure the area of a parallelogram as the area of a rectangle whose width is equal to the parallelogram's height and whose length is equal to the parallelogram's base can justify the result to themselves in the process. The process allows them to see

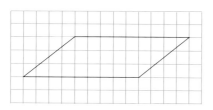

One way to justify the area formula for a parallelogram is to show how the area of a parallelogram can be reshaped as the area of a rectangle whose width is equal to the height of the parallelogram and whose length is equal to the base of the parallelogram.

Parallelogram	Base (cm)	Height (cm)	Area (cm²)
A	6	4	24
B	5	2	10
C	4	7	28
D	3	9	27

Fig. **2.4**.

A table constructed by students, showing values for the base, height, and area of four parallelograms

for themselves that the area of the rectangle, which they know is the product of its length and its width, is equal to the area of the parallelogram, which they can see is the product of its base and its height.

The parallelograms on the activity sheet can be decomposed and the parts easily recomposed as rectangles whose lengths and widths are equal, respectively, to the bases and heights of the given parallelograms. Many students will probably partition each parallelogram into a right trapezoid and a right triangle whose hypotenuse is one side of the original parallelogram (see fig. 2.1). Then the students can slide, or *translate*, the right triangle to the opposite side of the figure to form a rectangle whose length is equal to the base of the parallelogram, and whose width is equal to the parallelogram's height.

After the students have reshaped the areas of the given parallelograms as the areas of rectangles of equal area, they can begin to conjecture about whether and how they could reshape the area of *any* parallelogram in this way. Encourage your students to justify their conjectures by discussing why the product of base and height of a parallelogram is the area of each given parallelogram.

When the students have completed the activity sheet, select groups to share their results for the two given parallelograms. Once the class agrees on the area of each one, ask if any group obtained either measurement in a different way. Focus on the methods that the students used, encouraging them to generalize their work.

Although many students are likely to have reconfigured the areas of the parallelograms in the manner described above, others may have found different ways of decomposing the parallelogram and recomposing its parts as a rectangle whose length and width are equal to the parallelogram's base and height. For example, some may have cut off smaller right triangles from opposite sides of the parallelogram.

Figure 2.5 illustrates one such case. The hypotenuse of each smaller right triangle is half of the side of the parallelogram from which it is sliced; that is, the cuts that form these triangles pass through the midpoints of the parallelogram's sides. Ask students who have worked in this way, "Did you need to cut through the midpoints of the parallelogram's sides?" The students should see that cutting through the midpoints allows them to turn, or *rotate*, the triangles to recompose the parallelogram as a rectangle of equal area because the hypotenuse of

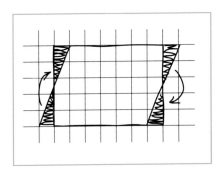

Fig. **2.5**.

Decomposing a parallelogram and recomposing it as a rectangle by slicing right triangles from opposite sides and rotating them

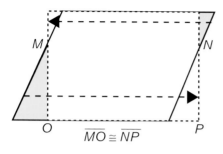

$$\overline{MO} \cong \overline{NP}$$

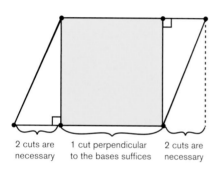

2 cuts are necessary | 1 cut perpendicular to the bases suffices | 2 cuts are necessary

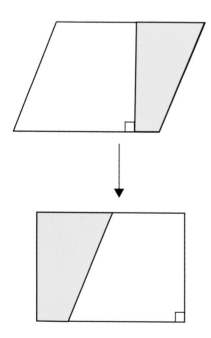

each new triangle is equal to the remaining, uncut part of the side of the parallelogram.

Other students may have cut right triangles from the sides of the parallelogram but not through the midpoints. These students may have discovered experimentally that these right triangles are similar though not congruent. The cuts that form them must slice through points that lie on opposite sides of the parallelogram at equal distances from one of the parallelogram's bases (see the margin). The students are then able to compose the rectangle by sliding the triangles to the opposite sides, rather than rotating them on the original sides, as described above.

Help your students understand that there are infinitely many ways to decompose the parallelogram and recompose the pieces as a rectangle whose base and height are the same as those of the parallelogram by making two cuts perpendicular to the bases. The sketch in the margin indicates the regions in which two cuts are necessary.

Likewise, the students can recognize that there are infinitely many ways to accomplish the decomposition and recomposition with just one such perpendicular cut. The shaded area in the sketch shows the region in which one cut perpendicular to the bases will decompose the parallelogram into pieces that can be recomposed as a rectangle whose length and width are equal, respectively, to the parallelogram's base and height. A single perpendicular cut that accomplishes this without passing through a vertex or the midpoint of a side is also illustrated in the margins, in the two sketches below on the left.

The length and width of the resulting rectangle are equal, respectively, to the base and height of the original parallelogram. Consider this rectangle in relation to the original parallelogram as shown in the sketch adjacent to the previous paragraph, divided into the regions necessitating one cut or two cuts. Note that the length of the resulting rectangle is equal to the length of the green (one-cut) region plus the length of one of the white (two-cut) regions. Imagine aligning the base of the newly composed rectangle with the base of the original parallelogram and sliding the rectangle all the way across the parallelogram from left to right (or right to left). The areas where the rectangle would "stick up" higher than the parallelogram would be the white areas, in which two perpendicular cuts would be necessary to decompose the parallelogram into pieces that could be recomposed as the rectangle in question. The region in which the "tops" of the rectangle and the parallelogram would align would be the green region, where just one cut would suffice.

As your students discuss their work, ask them if they think that their methods would work with any parallelogram. Could they consider any side of a parallelogram as the base and apply their method of decomposing the parallelogram to construct a rectangle of equal area with a length that is equal to that base? For example, could they use the short side of the parallelogram in figure 2.6a as the length of such a rectangle? Could they successfully apply their method in this case? They should find that they can make the required reshaping of the parallelogram's area, but they may need to modify or extend their method. Two ways of accomplishing the task are shown in parts (b) and (c) of the figure.

Before moving on to the next section of the investigation, you might discuss one last point with your students. Ask, "What is the relationship between a rectangle and a parallelogram?" Be sure that the students

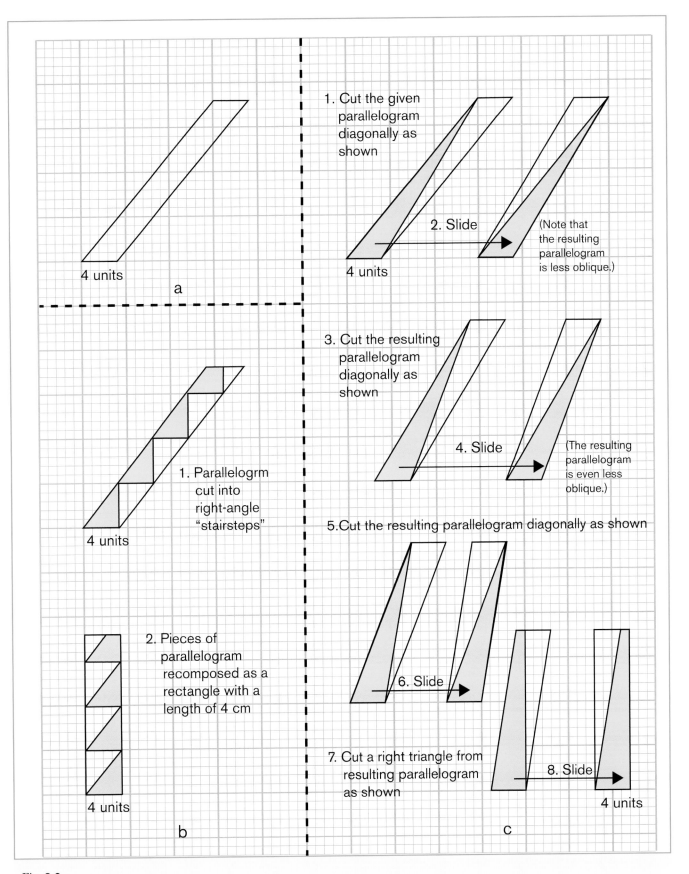

1. Cut the given parallelogram diagonally as shown

2. Slide

(Note that the resulting parallelogram is less oblique.)

4 units

4 units

a

3. Cut the resulting parallelogram diagonally as shown

4. Slide

(The resulting parallelogram is even less oblique.)

1. Parallelogrm cut into right-angle "stairsteps"

4 units

5. Cut the resulting parallelogram diagonally as shown

2. Pieces of parallelogram recomposed as a rectangle with a length of 4 cm

6. Slide

4 units

7. Cut a right triangle from resulting parallelogram as shown

8. Slide

4 units

b

c

Fig. **2.6.**

Can the 4-unit side of the parallelogram in (*a*) be used as its base in a decomposition that yields regions that can be recomposed as a rectangle whose length is 4 units and whose height is equal to that of the original parallelogram?

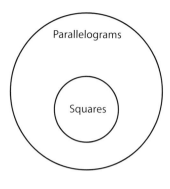

understand that all rectangles are parallelograms, but not all parallelograms are rectangles. A rectangle is a special case of a parallelogram—a parallelogram with four right angles.

To underscore the point, draw a circle on the board and tell your students that it represents the set of all parallelograms. Say, "If we want to show the set of all rectangles as another circle, where should we put that circle?" Suggest putting it entirely outside the circle, partly inside it, and completely inside it, letting the students consider each location. The students should understand that the second circle belongs inside the first one because the set of all rectangles is a subset of the set of all parallelograms.

Next, turn your students' attention to the familiar formula for the area of a rectangle (*Area = length × width*) and the formula that they have just derived for the area of a parallelogram (*Area = base × height*). Ask, "Are these formulas the same?" Let the students express the idea that they can use the new formula for *all* parallelograms—a set that includes all rectangles. Because all four angles of a rectangle are right angles, the side that the students think of as the rectangle's *width* actually is its *height*. Therefore, they can apply their new formula to rectangles as well as parallelograms that are not rectangles. By contrast, they can apply the old formula to all rectangles but not all parallelograms—only to parallelograms that are also rectangles.

Help the students understand that just as the definition of *parallelogram* is more general and inclusive than the definition of *rectangle*, so the formula for the area of a parallelogram is also more general and inclusive than the formula for the area of rectangle. They should recognize that they can use the new formula in place of the old one in determining the area of a rectangle simply by considering a rectangle as a special case of a parallelogram and thinking in terms of *base* and *height* instead of *length* and *width*.

Explore

Students can now apply the methods that they have used with parallelograms to explore the areas of triangles and trapezoids. Working with techniques of decomposition, transformation, and recomposition, along with their discoveries about the area of a parallelogram, they can derive the formulas for the areas of these new shapes.

Give each student a copy of the activity sheet "Moving into Triangular Territory," and direct the students to continue to work in groups. Ask, "Can you determine the area of a triangle by working as you did before and using what you learned about the area of a parallelogram?" If the students already know the formula for the area of a triangle, remind them that they are not to use it in their work: "Imagine that you have never heard of a formula for the area of a triangle."

When your students have completed the tasks on the sheet, ask several groups to share the methods that they used to determine the areas of the given triangles (see the margin). The students are likely to come up with many possible ways. Figure 2.7 shows two different ways in which eighth graders determined the area of the first triangle, which is a right triangle—specifically, an isosceles right triangle.

As your students share their methods for determining the area of a triangle, ask if these techniques would work with any triangle. How are

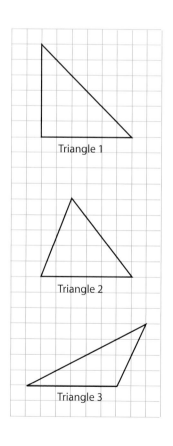

Make a cut that is perpendicular to the base of the right triangle and passes through the midpoint of its hypotenuse, producing a smaller right triangle (shaded gray) whose hypotenuse is one-half that of the original triangle. Rotate the new triangle into the position shown by the green triangle. Compute the area of the resulting rectangle.

Fig. **2.7**.

Determining the area of a right triangle by (*a*) decomposing the triangle and recomposing the parts as a rectangle and (*b*) making a rectangle that shares two sides with the triangle

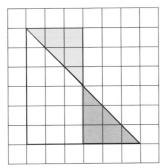

Area = 6 units × 3 units = 18 square units = height × $\frac{1}{2}$ base

a

Copy the right triangle and slide and rotate the copy (shown in green) to make a rectangle that shares two sides with the original triangle. Compute the area of the rectangle and divide by 2.

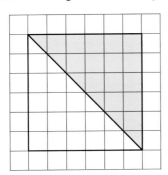

Area = (6 units × 6 units) ÷ 2 = (36 square units) ÷ 2 = 18 square units = height × base × $\frac{1}{2}$

b

the methods that they used with the first triangle, which is a right triangle, similar to or different from the methods that they used with the second and third triangle, which are, respectively, an acute triangle and an obtuse triangle?

Also be sure to discuss how the students' methods relate to the formula for the area of a triangle, often given as $A = \frac{1}{2}(b \times h)$. In particular, ask them to consider how the work that they did to determine the area of a triangle reflects the $\frac{1}{2}$ in the formula. For example, in the process shown in figure 2.7*a*, the base of the resulting rectangle is one-half the base of the original triangle, reflecting the division of the product of the triangle's base and the height by 2 in the formula for the area of a triangle. In the process shown in figure 2.7*b*, the rectangle composed of the original triangle and a copy has twice the area of the original triangle, pointing to the need to divide by 2, or multiply by $\frac{1}{2}$, to find the area of the original triangle.

Prepare your students to explore the area of a trapezoid by asking them to think about the methods that they have used so far, focusing on their similarities and differences. Comparing and contrasting methods are important reasoning processes. Ask which methods are similar and what the similarities are. Also ask if any method is different and how it differs. Students might offer descriptions like the following:

- "All the methods are similar because they change one shape into a different shape whose area we already knew how to find."
- "The method that we used to find the area of a parallelogram and the method that we used to find the area of a right triangle [illustrated in fig. 2.7*a*] are similar because they both cut and rearrange the shape to make a rectangle."
- "The method that we used to find the area of a right triangle [illustrated in fig. 2.7*b*] is different from others because with this method we made the rectangle by using a copy of the triangle along with the original triangle."

If time allows, have some students display their methods (see fig. 2.8) on poster board or chart paper. Post these displays in the classroom to summarize and reinforce learning from the activity.

Fig. **2.8.**

Students' representations of two methods: (*a*) decomposing and recomposing a shape to create a shape of equal area and (*b*) copying a shape and using the original and the copy to compose a shape with twice the area of the original

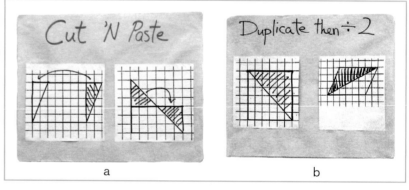

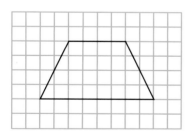

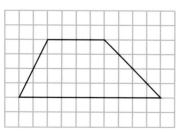

Have the students continue exploring, this time with trapezoids. Give each student a copy of the activity sheet "How Much Area Does a Trapezoid Trap?" Step 1 presents two trapezoids (see the margin), the first of which is isosceles. The students must determine the area of each trapezoid by at least two different methods. Urge them to think about their earlier work with parallelograms and triangles and reflect on the discussion in which they summarized the methods that they used in those cases.

Step 2 asks the students to explain whether their methods would allow them to find the area of any trapezoid. A formula is a generalization that depends on the identification of a method that works in every case. After the students have explained a method for the general case, they move on to step 3, which asks them to apply their method to determine the areas of trapezoids that are not shown on grids but have measurements specified for their bases and heights. When students attempt to find the area of a figure on a grid, they may simply count grid squares. Working at this stage with trapezoids without grids can help the students focus on what happens to particular measurements when they manipulate the figures.

At this point, the students should be considering what measurements they need to know to determine the area of a trapezoid. Step 3 helps them think about this question. An alternative way to motivate them to think about the necessary measurements is to give them a trapezoid on unlined paper. Specify no measurements at all, and ask them to work in their groups to determine which measurements (no more than four in all) they must know to be able to determine the trapezoid's area. After the groups have considered this question, ask them to share their ideas and strategies for using those measurements to obtain the area. This experience can prepare them for step 4, which asks them for a statement of the formula.

Evaluate

Discuss your students' work with trapezoids in detail. Consider their comments carefully, since these can give you a good indication not only of how well they understand the area of a trapezoid but also of how well they understood their earlier work with the areas of a parallelogram and a triangle. As the students share the methods that they used with trapezoids, encourage them to relate these methods to those that they used with the earlier shapes.

In the class discussion, your students are likely to present a variety of methods to determine the area of a trapezoid. Thus, while students are listening to their classmates' explanations, they should be thinking about how each method is similar to or different from their own. When a student or group of students shares a method with the class and explains whether or not it applies to any trapezoid, draw an arbitrary trapezoid on the board, and ask for a demonstration of the method on it.

The first trapezoid in step 1 is isosceles, so some methods that your students are likely to have used to find its area will not work with all trapezoids. Figure 2.9 shows two methods that work with an isosceles trapezoid but will not be successful with other trapezoids. Figure 2.9*a* shows a partitioning of the trapezoid into a right trapezoid and a right triangle. Sliding and flipping (*translating* and *reflecting*) the right triangle as shown produces a rectangle of equal area. Figure 2.9*b* shows a partitioning of the trapezoid into two parts along its line of symmetry, producing two right trapezoids that are mirror images of each other. Sliding and flipping one of these new trapezoids as shown produces a rectangle that is congruent to that shown in (*a*).

In fact, *any* cut that passes at a right angle through both bases of an isosceles trapezoid decomposes the trapezoid into two regions that can be recomposed as a rectangle of equal area. This point is worth making, even though the geometry involved in the explanation is perhaps too advanced for most middle-grades students. Such a partitioning of an isosceles trapezoid results in two regions that consist either of a right triangle and a right trapezoid or of two right trapezoids. In the two cases where the partitioning of an isosceles trapezoid *ABCD* produces a right triangle (see the upper figure in the margin on the next page), angles *ABC* and *BAC* are complementary. Because the trapezoid is isosceles, angles *ABC* and *CDE* are congruent, so angles *BAC* and *CDE* are also complementary. When the right triangle is flipped and slid in the

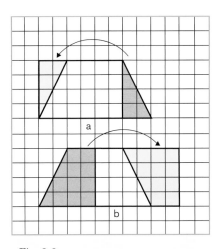

Fig. **2.9.**

Two methods that help in finding the area of an isosceles trapezoid but do not work with trapezoids that are not isosceles

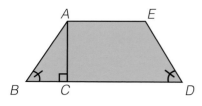

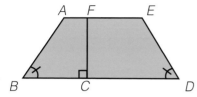

manner suggested in figure 2.9a, angles *BAC* and *CDE* merge to form a right angle, and as a result the two parts of the original trapezoid are recomposed as a rectangle.

The case of the infinite number of partitions that produce two right trapezoids is similar. See the sample partitioning of trapezoid *ABDE* in the lower margin figure. Angles *ABC* and *FAB* are supplementary, as they would be in any trapezoid, isosceles or not:

$$m\angle ABC + m\angle BCF + m\angle CFA + m\angle FAB = 360^\circ.$$
$$\text{But } m\angle BCF + m\angle CFA = 180^\circ.$$
$$\therefore m\angle ABC + m\angle FAB = 180^\circ.$$

However, because trapezoid *ABDE* is isosceles, angles *ABC* and *CDE* are congruent, and thus angle *FAB* is supplementary to angle *CDE* as well as angle *ABC*. The original isosceles trapezoid is partitioned by a perpendicular cut through the bases, so angles *AFC* and *FCB* are right angles. When the parts of the decomposed trapezoid are recomposed as suggested in figure 2.9b, the supplemental angles come together to form line segments in a figure that is a quadrilateral with four right angles—a rectangle. Clearly, the method of making a perpendicular cut through the bases to decompose a trapezoid and recompose it as a rectangle of equal area depends on starting with a trapezoid that is isosceles.

The second trapezoid that the students encounter on the activity sheet is not isosceles. Figure 2.10 shows two methods that some eighth graders used to determine its area. The first method, pictured in figure 2.10a, extends to trapezoids the method of copying, sliding, and rotating the original shape that some students probably used with triangles (see fig. 2.7b). As in the earlier case, the resulting shape is a parallelogram with twice the area of original shape. The students can find the parallelogram's area easily and divide by two to determine the area of the trapezoid or triangle in question.

Fig. **2.10.**

Two methods that students used to determine the area of a trapezoid that is not isosceles

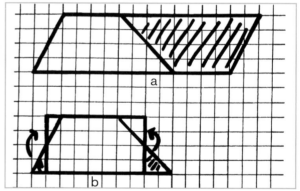

The second method, pictured in figure 2.10b, approaches the problem by "evening up" the trapezoid's parallel sides to create a rectangle of equal area. This work extends to trapezoids the method of slicing and rotating that some of your students probably devised in working with triangles (illustrated in figure 2.7a). As in the case of the triangle, each small right triangle that is cut from the trapezoid has a hypotenuse that is half as long as the side of the original side from which it came; that is, once again the cut passes through the midpoint of the side.

In discussing this method with your students, ask a related question: "If a trapezoid does not appear on a grid but you know the lengths of its bases, could you determine the length of a rectangle with the same area?" The trapezoids that the students encounter in step 3 of the activity sheet are presented in this manner—without grids but with measurements specified for the bases.

Two students in one classroom presented their different methods for determining the length of the rectangle with the same area for the trapezoid that the activity sheet presents in step 3(*a*). The trapezoid is shown in the margin.

Figure 2.11 illustrates the first student's method. This student began by cutting off two small right triangles, shown in gray in figure 2.11*a*. The student discussed how to rotate these triangles into the positions shown by the green triangles, to form the rectangle, as indicated in the figure. Then the student drew the larger right triangles labeled as *ABF* and *CDG* in figure 2.11*b* and offered the following explanation of his process for determining the length of the rectangle in question:

> If you draw perpendicular lines to make two triangles [*ABF* and *CDG* in fig. 2.11b], you see that the total of the extra lengths [*AB* and *CD* in the figure] is 8 units. Because each of these [large right] triangles is cut at the midpoints of their bases to form the new rectangle, the portion of the length of the extra segments added on to the top of the original trapezoid [*EF* and *GH*] should be 4 units. Thus, the new base should be 10 units long.

This student's ability to work with the two unknown lengths by knowing that the total of those two lengths is 8 units reveals an algebraic component in his reasoning.

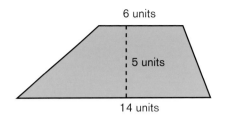

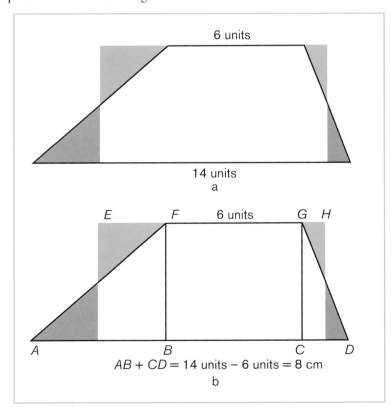

Fig. **2.11.**

One student's method for determining the length of a rectangle whose area and height are equal to those of a trapezoid whose bases are 6 units and 14 units

The second student applied her general understanding of average (or mean) to the new situation, using a familiar concept to construct a different but equally valid justification for 10 units as the length of the rectangle with a height and area equal to those of the given trapezoid:

What I am doing is trying to "even out" the two bases. I know that when we "even out" two (or more) quantities, we are finding the average (mean) of those quantities. So, the length of the new base is $\frac{6+14}{2}$ units.

Your students may devise many different methods to manipulate or duplicate the area of a trapezoid so that they can determine its measure from what they know about the area of a rectangle, a parallelogram, or a triangle. No matter what method they develop, they should write an expression for the formula that closely matches their method. For example, suppose that students use the method depicted in figure 2.10*a*, which shows a trapezoid that is copied and the copy is translated and rotated to form a parallelogram with twice the area of the trapezoid. The students' formula should look something like the following:

Area = [(long base + short base) × height] ÷ 2.

Suppose instead that students use the method shown in figure 2.10*b*, in which small right triangles are cut from the trapezoid and rotated to reconfigure its area as that of a rectangle. In this case, students should write a different expression—at least before simplifying—as a formula for the area of a trapezoid. Depending on how they determine the width of the new rectangle, they might write either of the following:

Area = {[(long base – short base) ÷ 2] + short base} × height

or

Area = [(long base + short base) ÷ 2] × height.

Other methods would give rise to a variety of corresponding expressions of the formula. Figure 2.12 shows a number of additional methods of reconfiguring the area of a trapezoid as the area of a rectangle, a parallelogram, a triangle, or a combination of these shapes. The figure pairs each method with a corresponding expression of the formula in unsimplified form.

The method pictured in figure 2.12*a* makes a cut that is parallel to the trapezoid's bases and passes through the midpoints of the trapezoid's other sides. This cut creates two new trapezoids with equal height (each of which is one-half of the original trapezoid's height), unequal area, and a shared base. Rotating either of these new trapezoids produces a parallelogram whose area is equal to that of the original trapezoid and whose base is equal to the sum of the trapezoid's long and short bases. The expression

Area = (long base + short base) × (height ÷ 2)

reflects the method of decomposing and recomposing the area.

Figure 2.12*b* shows a cut from a vertex of one base of the trapezoid to the diagonally opposite vertex of the other base. This cut produces two triangles of equal height (each of which is the same height as the original trapezoid), a shared side, and unequal area. The long base of the trapezoid forms the base for one of the triangles, and the trapezoid's short base serves as the base of the other triangle. Thus, the expression

Area = $\frac{1}{2}$(long base × height) + $\frac{1}{2}$(short base × height)

represents the process of creating the two triangles from the trapezoid.

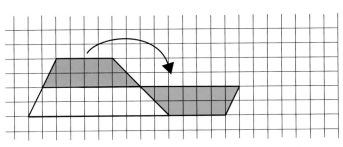

Area = (long base + short base) × (height ÷ 2)

a

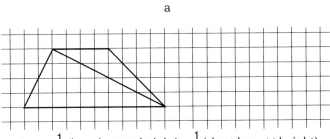

Area = $\frac{1}{2}$ (long base × height) + $\frac{1}{2}$ (short base × height)

b

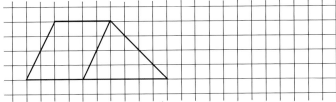

Area = (short base × height) + $\frac{1}{2}$ (long base – short base) × height

c

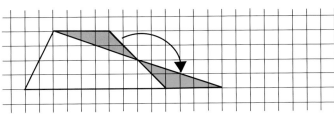

Area = $\frac{1}{2}$ (long base + short base) × height

d

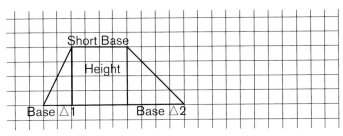

Area = (short base × height) + $\frac{1}{2}$(base △1 × height)
$+ \frac{1}{2}$(base △2 × height)

e

Fig. 2.12.

Other methods of reconfiguring the area of a trapezoid as the area of a rectangle, a parallelogram, a triangle, or a combination of these shapes

The cut in figure 2.12c slices through a vertex that joins the trapezoid's short base and one of its nonparallel sides, making a line that is parallel to the trapezoid's other nonparallel side. Thus, the cut divides the trapezoid into a parallelogram and a triangle of equal height. The expression

$$\text{Area} = \text{short base} \times \text{height} + \tfrac{1}{2}(\text{long base} - \text{short base}) \times \text{height}$$

directly reflects the decomposition of the trapezoid into a parallelogram and a triangle and the calculation of their areas with known formulas.

Figure 2.12d pictures a cut and a rotation that reshape area of the trapezoid as the area of a triangle. As in figure 2.12c, the cut slices through a vertex on the trapezoid's short base and the midpoint of the trapezoid's opposite nonparallel side. The triangle made by the cut can be rotated as shown, recomposing the trapezoid as a single triangle of equal area. The height of the new, large triangle is equal to that of the original trapezoid, and its base is equal to the sum of the trapezoid's bases. The expression

$$\text{Area} = \tfrac{1}{2}(\text{long base} + \text{short base}) \times \text{height}$$

reflects this method of creating and finding the area of the triangle of equal area.

The method in figure 2.12e makes the short base of the trapezoid the side of a rectangle formed by two perpendicular cuts through the base's vertices. (The rectangle happens to be a square in this example.) These cuts produce two triangles in addition to the rectangle. Note that the corresponding expression in figure 2.12e,

$$\text{Area} = (\text{short base} \times \text{height}) + \tfrac{1}{2}(\text{base } \triangle 1 \times \text{height})$$
$$+ \tfrac{1}{2}(\text{base } \triangle 2 \times \text{height}),$$

suggests that this method requires measurements of four different lengths—the trapezoid's short base, its height, and the base of each newly created triangle—for a determination of the area of the rectangle and each triangle.

By contrast, the other methods discussed so far have called for measurements of three lengths—the trapezoid's long base, its short base, and its height. However, students can readily modify the method in figure 2.12e by adding a step. This step, consisting of two translations (see the margin), recomposes the two right triangles as a single triangle whose height is equal to that of the original trapezoid and whose base is equal to the trapezoid's long base minus its short base. As modified, the method thus reconfigures the original trapezoid as a triangle and a rectangle that do not share a side but whose areas add up to the area of the original trapezoid. The corresponding expression would need only the same three measurements as the other methods and would be identical to that for the method in figure 2.12c, which shows the decomposition of the trapezoid into a parallelogram and a triangle:

$$\text{Area} = \text{short base} \times \text{height} + \tfrac{1}{2}(\text{long base} - \text{short base}) \times \text{height}$$

Note that the triangle produced by the new step is congruent to the triangle in figure 2.12c, and the rectangle has the same height and base,

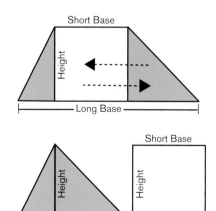

and thus the same area, as the parallelogram produced by the earlier method.

As your students discuss their expressions of the formula for the area of a trapezoid, make sure that each expression matches the method. This can be especially important if students are already familiar with the formula for determining the area of a trapezoid. The closeness of the match between the expression and the method can be a good indicator of their understanding of how and why the formula works. For students who are familiar with algebra, another useful exercise can be determining whether or not the various expressions that they have come up with are equivalent.

Extend

Students can readily extend their work in the investigation by exploring ways of determining the areas of two additional shapes: a kite and a "stair" (see the margin). Formulas for the areas of these shapes are not usually discussed in mathematics classes, so developing them should afford new experiences to most students.

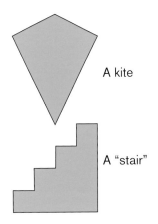

A kite

A "stair"

A kite is a quadrilateral with two pairs of adjacent congruent sides and exactly one line of symmetry. Thus, a rhombus, which has two pairs of adjacent congruent sides but two lines of symmetry, is not a kite. The diagonals of a kite are perpendicular to each other and the point at which they intersect is the midpoint of one diagonal but not the other.

The activity sheet "Carryover to Kites" presents two kites for exploration. On the basis of their earlier work in the investigation, your students can manipulate a kite in a number of different ways to determine its area and then give formulas that reflect their work. Figure 2.13 shows four possible ways of decomposing the kite presented as kite 2 on the activity sheet and preserving its area in triangles or rectangles whose areas the students can calculate easily with formulas that they now know.

Figure 2.13a shows a cut through the kite's line of symmetry (the long diagonal of the kite shown), decomposing the shape into two triangles that are reflections of each other. Using the formula $Area = \frac{b \times h}{2}$ for the area of each triangle, the students obtain a measurement of $\frac{8 \times 3}{2}$, or 12 square units for each. Thus, the area of the kite is 24 square units.

Figure 2.13b shows a cut through the kite's short diagonal (for convenience, D_1), decomposing the shape into two different triangles. These triangles share D_1 as their base. Using the formula $Area = \frac{b \times h}{2}$ for the area of each triangle, the students can compute the smaller triangle's area as $\frac{6 \times 2}{2}$, or 6 square units, and the larger triangle's area as $\frac{6 \times 6}{2}$, or 18 square units. Adding 6 and 18, they again obtain 24 square units as the area of the kite.

The work illustrated in figure 2.13c shows cuts through both diagonals, decomposing the kite into four right triangles. Each triangle is then copied, and each copy is rotated in such a way that the copies and the original triangles together compose a rectangle with twice the area of the kite. (The students know that the rectangle's area is two times that of the kite because each new (shaded green) triangle is congruent to the unshaded triangle that is adjacent to it but inside the kite.) By applying the formula for the area of a rectangle and dividing the result by 2, the students determine that the area of the kite as $\frac{8 \times 6}{2}$, or 24, square units.

The method shown in figure 2.13*d* shows a cut all the way through the long diagonal (D_2) and halfway through the short diagonal (D_1), from one vertex to the diagonal's midpoint, which is the point at which D_1 and D_2 intersect. These two cuts decompose the kite into three triangles, two of which are to the right of D_2 and shaded gray in figure 2.13*d*. Flipping, rotating, and sliding these shaded triangles into new positions (shown by the green triangles) adjacent to the unshaded triangle on the left allows the creation of the rectangle on the left-hand side of the figure. The area of this rectangle is the same as that of the kite. Students can easily compute the rectangle's area: 3 × 8 = 24 square units.

Fig. **2.13.**

Four ways of using decomposition to determine the area of kite 2: (*a* and *b*) cutting it into two triangles; (*c*) cutting it into four triangles, copying them, and building a rectangle of twice the area; and (*d*) cutting it into three triangles and recomposing them as a rectangle of equal area

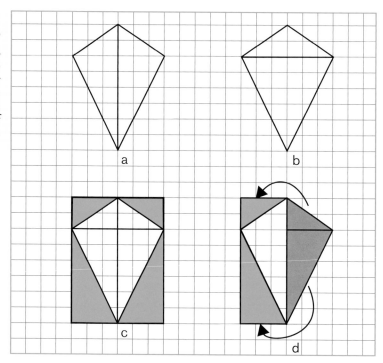

After investigating the four methods shown, students may wonder how to use the lengths of the two diagonals, D_1 and D_2, to help them find the area of a kite. Figure 2.13*c* highlights how this process works. When the students construct the surrounding rectangle whose dimensions are D_1 by D_2, they know that the area of the kite is half the area of the rectangle, so *Area* $= \dfrac{D_1 \times D_2}{2}$.

Investigating the area of a "stair" offers another way for the students to extend their understanding of the areas of parallelograms, triangles, and trapezoids. The activity sheet "Stepping Up to Stairs" presents a stair as a two-dimensional shape consisting of a series of adjacent rectangles. One dimension of the rectangles remains constant while the other dimension grows by a constant amount from one rectangle to the next.

A real staircase of course has height and a baseline; in talking about one of the stairs on the activity sheet, the students thus may find it more natural to characterize a rectangle's dimensions as *base* and *height*, rather than as *length* and *width*. Likewise, they may find it more straightforward and less confusing to work with the area formula *Area = base × height*, which applies to all parallelograms, including the special case of rectangles, rather than the formula *Area = length × width*.

The students can think of the adjacent rectangles in a stair as resting on same-sized bases that "line up" on the "floor" or "ground" from which the stair rises. Because each rectangle has a greater height than the previous one, each rectangle except the "highest" has one side that is completely contained in a side of the next rectangle.

Note that the height of the "first," or "shortest" rectangle need not be equal to the constant difference between the heights of the adjacent rectangles in the stair. See, for example, the stair in the margin, which appears as stair 2 on the activity sheet. In this case, the first rectangle is 3 units high, but the constant difference between the heights of successive rectangles in the stair is 2 units. Thus, stepping onto the first step to go up the stair would involve making a bigger step than would be necessary in stepping from one step to the next after that. In other cases, stepping onto the stair could involve making a smaller step than stepping from one step to the next would require from then on. (In either case, someone coming down the stair might well find that "that last step is a dilly," as the old saying goes!)

Figure 2.14 shows three possible ways in which students might determine the area of the shape presented as stair 1 on the activity sheet. These processes enable them to think of the area of the stair in terms of the area of a rectangle by one of two methods—either by making an identical copy of the stair and rotating it to form a large rectangle with twice the area of the stair (fig. 2.14*a* and fig. 2.14*b*) or by cutting, sliding, and rotating a part of the stair (fig. 2.14*c*) to make a single rectangle that is half as high as the stair and as long as the stair plus the length of one additional step.

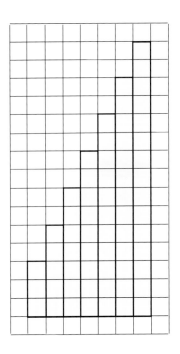

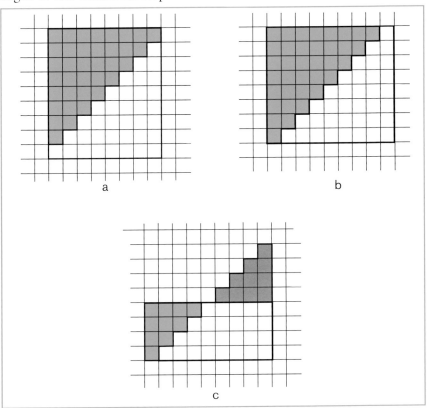

Fig. **2.14.**

Three ways to manipulate stair 1 and determine its area as the area of a rectangle

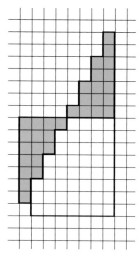

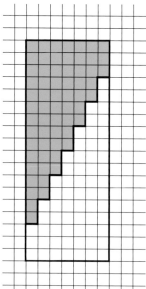

The students should note that they cannot apply the second method —the method that produces the "shallower" rectangle of equal area— to the second stair on the activity sheet. The method does not produce a rectangle in this case because of the disparity between the height of the stair's first rectangle and the constant difference in the rectangles' heights (see the margin). Also note that in the case of a stair with an odd number of steps, with or without such a disparity, applying the second method is a little different. It involves halving the height of the middle step, which eliminates the need for a slide. To create the rectangle with half the height, a simple rotation is sufficient.

However, the first method suggested above—the method that duplicates the stair and uses the copy to complete a single large rectangle with twice the area (see the margin)—can help the students write an equation that they could use to determine the area of *any* stair. Coming up with such an equation is an important outcome of the activity.

Using this method can lead to the development of a formula like the following for the area of a stair:

$$\text{Area} = \frac{1}{2}[(\text{number of rectangles} \times \text{base of each}) \times (\text{height of first rectangle} + \text{height of last rectangle})].$$

The number of steps in a stair is equal to its number of adjacent rectangles. The distance from the "front" to the "back" of each step is the same and is equal to the number of units in the base of each rectangle. The number of rectangles times the number of units in the base gives the number of units in the base of the large rectangle formed by the green and white regions in figure 2.15. Furthermore, the height of the first ("shortest") adjacent rectangle plus the height of the last

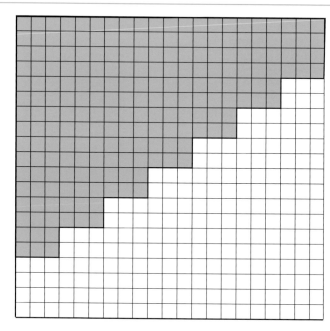

Fig. **2.15.**

Copying a stair and rotating the copy to make a rectangle with twice the area before applying the formula *Area = base × height* and dividing by 2 to determine the area of the stair

$\text{Area} = \frac{1}{2}[(\text{number of steps} \times \text{"length" of a step}) \times (\text{height of first step} + \text{height of last step})]$

$= \frac{1}{2}[(7 \times 3 \text{ units}) \times (4 \text{ units} + 16 \text{ units})]$

$= \frac{1}{2}(21 \text{ units} \times 20 \text{ units})$

$= 210 \text{ square units}$

Navigating through Problem Solving and Reasoning in Grades 6–8

("highest") adjacent rectangle gives the height of this large rectangle. Substituting the numbers of units in the base and height of the large rectangle into the formula *Area = base × height* gives this rectangle's area, and dividing by 2 gives the area of the stair.

This process of determining the area of a stair prepares students for later work in finding the sum of an arithmetic sequence—a sequence with a constant difference. Note the similarity between the formula above and that for the sum of the first *n* terms of an arithmetic sequence:

$$\text{Sum of first } n \text{ terms} = \tfrac{1}{2}\,[n \times (\text{sum of first and } n\text{th terms})].$$

Conclusion

Reasoning plays an important role throughout the study of measurement. This chapter has illustrated the role of reasoning in the development of middle-grades students' understanding of important concepts related to the measurement of area. The activities in the investigation presented in the chapter have focused on the reasoning involved in the process of deriving formulas for the areas of fundamental two-dimensional shapes—parallelograms, triangles, and trapezoids.

NCTM's Illuminations Web site features a number of resources with embedded applets that provide varied interactive opportunities for students to explore the areas of triangles, trapezoids, and parallelograms. Several of these resources are identified in the margin.

The students' work in the investigation emphasizes the fact that a formula is a generalization whose derivation necessarily requires looking for patterns and relationships and making conjectures and evaluating them. The next chapter shows the role of reasoning in the development of students' understanding of ideas in a closely related area of mathematics—geometry.

The work that students have done with figures on grids in this investigation can also serve as a foundation on which to build or extend an understanding of the Cartesian coordinate system. The accompanying CD-ROM includes the applet Triangle Explorer, which helps familiarize students with this two-dimensional mapping system as they determine the areas of triangles drawn on a Cartesian grid.

"When students are presented with exploratory measurement environments rather than proof-based environments, they can build their thinking and reasoning skills, which in turn can help in the development of mathematical reasoning." (Enderson 2003, p. 271; available on the CD-ROM).

The Illuminations Web site offers opportunities for students to explore the areas of rectangles, parallelograms, triangles, and trapezoids interactively:

- Interactive Geometry Dictionary: Areas in Geometry
  http://illuminations.nctm.org/ActivityDetail.aspx?ID=21

- Discovering the Area Formula for Triangles
  http://illuminations.nctm.org/LessonDetail.aspx?ID=L577

- Finding the Area of Trapezoids
  http://illuminations.nctm.org/LessonDetail.aspx?ID=L580

NAVIGATIONS SERIES

GRADES 6-8

PROBLEM SOLVING *and* REASONING

Chapter 3
Reasoning about Geometry

"By reflecting on [problems and] their solutions…, students use a variety of mathematical skills, develop a deeper insight into the structure of mathematics, and gain a disposition toward generalizing. … It should become second nature for students to talk about connections among problems; to propose, critique, and … be adept in explaining their approaches."

(NCTM 2000, p. 261)

Mathematical understanding in middle school builds on students' experiences in finding geometric patterns and describing relationships. The students formulate and test conjectures as they create sound and increasingly complex arguments, though these usually lack the precision and rigor of formal proofs. In addition, discussions—both among students and between students and teachers—provide students with opportunities to evaluate their own thinking as well as that of their peers. Classrooms that promote inquiry and communication provide essential experiences that build proficiency in using inductive and deductive reasoning.

Reasoning Experiences in Geometry

Reasoning about geometric concepts extends middle school students' thinking. Opportunities to engage in geometric reasoning build on their informal experiences in searching for patterns to explain the world. This reasoning is inductive, and its development is important in fostering the more formal, deductive reasoning that becomes increasingly significant in the study of geometry at the secondary and postsecondary levels.

Inductive reasoning involves students in making observations, discovering patterns, formulating tentative conjectures, and testing them before making generalizations. For example, students who draw the diagonals of various rectangles and record the diagonals' lengths may notice that each rectangle's diagonals are congruent. They might then

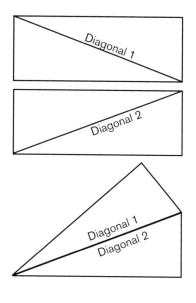

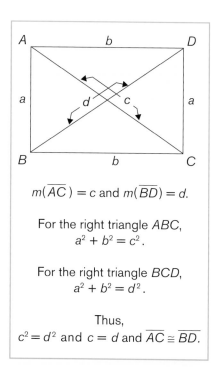

$m(\overline{AC}) = c$ and $m(\overline{BD}) = d$.

For the right triangle *ABC*,
$a^2 + b^2 = c^2$.

For the right triangle *BCD*,
$a^2 + b^2 = d^2$.

Thus,
$c^2 = d^2$ and $c = d$ and $\overline{AC} \cong \overline{BD}$.

Fig. **3.1.**

Using the Pythagorean theorem to demonstrate deductively that the diagonals of a rectangle are congruent

Navigating through Geometry in Grades 6–8 (Pugalee et al. 2002) emphasizes the application of geometric reasoning in a variety of problem-solving contexts.

conjecture that the diagonals of any rectangle are congruent. Such a conjecture results from reasoning inductively on the basis of observations about events, patterns, and relationships.

How might middle school students justify this conjecture about the diagonals of a rectangle? They might make a copy of a rectangle and draw one diagonal on the original and the other on the copy (see the margin). They could cut out both rectangles and divide both on the marked diagonals. Then they could "match up" the diagonals to show that they are congruent.

Alternatively, they might use the Pythagorean theorem to show deductively—using algebra—that the diagonals are equal (see fig. 3.1). To apply the theorem, they would work from the definitions of a rectangle as a quadrilateral with four right angles and of a right triangle as a triangle with a right angle. Because rectangle *ABCD* has four right angles, sides *AB* and *CD* are congruent (both with length *a*), and sides *BC* and *AD* are also congruent (both with length *b*). Because △*ABC* and △*BCD* are right triangles, the diagonals of rectangle *ABCD* (*AC* and *BD*) are the hypotenuses of different right triangles. A straightforward application of the Pythagorean theorem shows that the lengths of the diagonals are equal.

Deductive reasoning, a vital foundation for mathematics, is the method by which conclusions are drawn in geometric proofs. Such reasoning is used to show the logical certainty of a principle.

The Development of Geometric Reasoning

The work of two Dutch educators, Pierre van Heile and Dina van Heile–Geldof, has shed light on how students develop geometric reasoning. This valuable research identifies levels of thinking and understanding that are determined in part by the objects of geometric thought. As students move from a lower to a higher level, these objects change. Although the van Hiele levels do not correspond specifically to grade levels, they offer a useful way to characterize students' reasoning at various developmental levels. The following descriptions are based on those provided by Van de Walle (2007).

Level 0–Visualization

Students at level 0 can name and recognize shapes by their appearance but cannot identify their specific properties. Such students recognize, for example, that triangles and squares have different numbers of sides but are unaware that the sizes of the angles matter or that changes in orientation do not change the shapes. Reasoning at this level is most evident in students in the elementary grades.

Level 1–Analysis

At level 1, students begin to identify properties of shapes. They distinguish unchanging information (for example, that a square has congruent sides, right angles, and perpendicular diagonals) from changing information (for example, the size of a square [is it small or large?] and its orientation in space). They begin to understand that if a shape belongs to a class like "square," then it has all the properties of that class. Reasoning at this level generally develops by late elementary school or, for some, in middle school.

Level 2–Informal deduction

Students who reason at level 2 can identify relationships between and among properties of shapes or classes of shapes and can organize shapes into hierarchies on the basis of their properties. For example, they clearly understand that squares are rectangles because all four angles of a square are right angles. They appreciate logical arguments and engage in "if ... then" reasoning, but their proofs are usually more intuitive than rigorously deductive. Level 2 reasoning generally develops in middle school, although some students do not reason at this level until high school.

Level 3–Deduction

Students who reason at level 3 consider relationships among axioms, definitions, theorems, corollaries, properties, and postulates and use them in constructing proofs. High school geometry courses typically expect reasoning at this level, which prepares students for subsequent work in different geometric or axiomatic systems.

Level 4–Rigor

At level 4—the highest level in the hierarchy—the focus is on reasoning about axiomatic systems themselves—not simply on reasoning to make deductions within a system. A college mathematics major who is studying geometry as a branch of mathematics is generally able to reason at this level.

Reasoning and Geometric Relationships in the Middle Grades

Understanding geometric relationships is at the heart of middle school mathematics. As *Principles and Standards for School Mathematics* (NCTM 2000) asserts, students typically come to middle school with informal knowledge about "points, lines, planes, and a variety of two- and three-dimensional shapes" (p. 233)—knowledge that they have acquired by manipulating many objects, in and out of school. They probably have had numerous experiences in "visualizing and drawing lines, angles, triangles, and other polygons," preparing them to "investigate relationships by drawing, measuring, visualizing, comparing, transforming, and classifying geometric objects" (p. 233) in middle school. At this stage, investigations in geometry can provide them with contexts for "developing mathematical reasoning, including inductive and deductive reasoning, making and validating conjectures, and classifying and defining geometric objects" (p. 233).

"To prepare students for more formal thinking in the secondary school, geometry activities in the middle school must go beyond simple visual exercises. Rigorous proofs are not necessary at this age, but students should be able to use ideas about geometry to construct informal arguments, which helps them better understand the structure of geometry."
(Carroll 1998, p. 403)

Malloy (2002; available on the CD-ROM) describes the van Hiele levels in more detail.

Problems that lead middle-grades students to explore geometric relationships can help them make strong connections among geometric concepts and other topics in mathematics while strengthening their skills in reasoning, problem solving, and communicating. The following investigation presents a problem that demonstrates how connections can become explicit in explorations of geometric relationships. The problem in the investigation offers multiple entry points for students at different levels, and the discussion of it in the text is rooted in work with several middle-grades mathematics classes.

Reasoning about Similar Figures

Goals

- Work with a geometric task whose cognitive demand is consistent with reasoning at level 2 in the van Hiele framework (identifying relationships among properties of shapes or classes of shapes and organizing shapes according to their properties)
- Extend skills in reasoning about geometric relationships
- Construct a sound mathematical argument about a geometric relationship
- Expand concepts related to congruence, similarity, symmetry, and dilation of a shape
- Reinforce and extend problem-solving skills through making, examining, and defending conjectures about geometric relationships
- Support the goals of the Communication Standard (NCTM 2000, p. 268):
  - Organize and consolidate … mathematical thinking through communication
  - Communicate … mathematical thinking coherently and clearly to peers, teachers, and others
  - Analyze and evaluate the mathematical thinking and strategies of others
  - Use the language of mathematics to express mathematical ideas precisely

Materials and Equipment

For each student—

- A copy of each of the following activity sheets:
  - "Diagonal Discoveries"
  - "Rectangles for 'Diagonal Discoveries'"
  - One or two sheets of centimeter grid paper (template on the CD-ROM)

For each group of two to four students—

- A pair of scissors
- A ruler (calibrated in metric units)

Prior Knowledge

Experiences in grades 3–5 should have given students numerous experiences in working with rectangles, including congruent rectangles, similar rectangles, and rectangles with different dimensions and proportions. Such experiences, along with work with other shapes, should have introduced them to the concepts of similarity and congruence. They

The problem in Reasoning about Similar Figures is an example of a task that makes a cognitive demand on students that is consistent with reasoning at level 2 in the van Hiele framework.

pp. 154–55, 156–158

Use the template "Centimeter Grid Paper," which appears on the CD-ROM, to print sheets of grid paper for your students' use in the investigation Reasoning about Similar Figures.

should be acquainted with the idea that two shapes are congruent if they have exactly the same shape and size (all corresponding angles and all corresponding sides are congruent) and that two shapes are similar if they have corresponding angles that are congruent and corresponding sides that are in the same ratio.

Learning Environment

The students work on the investigation in pairs or groups of three, with the teacher serving as a facilitator, listening carefully and asking questions to help them formulate conjectures and construct justifications. When the students complete the activity, they share their conclusions and supporting arguments with the class. The teacher prompts them to revisit ideas that lack sound mathematical reasoning or need additional support.

Discussion

The fundamental mathematical concept underlying the problem in this investigation is *similarity*. Middle school students' explorations of similarity complement their investigations of proportional reasoning (ratio, fractions, percentages, etc.) in the areas of number and algebra. In geometry, students need many opportunities to explore the properties of similar figures and understand relationships among their angles, side lengths, perimeters, and areas. Work with similarity helps students consolidate their understanding of the related concept of *congruence* as they distinguish figures that are congruent from those that are similar. In the process, they should create and critique arguments that involve these two concepts, as well as ideas related to *symmetry*.

Middle school students in different grades can reason about and extend their understanding of similarity by applying ideas about similar figures to coordinate geometry and by working with transformations known as *dilations*. A dilation yields an image that is the same shape as the original figure but on a different scale.

This exercise involves students in developing and using habits of mind identified by Cuoco, Goldenberg, and Mark (1996) as essential for success in mathematics. These include visualizing, experimenting, describing with informal and formal language, making and using definitions, finding invariants, conjecturing, and justifying. Sound reasoning builds students' understanding of why something occurs and how it connects with other mathematical ideas and principles that they have encountered previously.

Engage

In a class discussion, elicit what your students know about rectangles by asking, "What makes a rectangle a rectangle?" Urge the students to brainstorm quickly, and as they offer their ideas, record them on the board. Write their suggestions without comment, including characteristics that do not uniquely identify a quadrilateral as a rectangle, such as the following, which are properties of all parallelograms:

- "The lengths of the opposite sides are equal."
- "The opposite angles are congruent."

For an example of an activity that introduces dilations at the middle school level, see Dilating Figures in *Navigating through Geometry in Grades 6–8* (Pugalee et al. 2002, pp. 21–22, 92).

- "A diagonal forms two congruent triangles."
- "The opposite sides are parallel."
- "The diagonals bisect each other."

Let the students sort through all the characteristics that they suggest, identifying those that do belong uniquely to a rectangle, such as the following:

- "All the angles are right angles (90°)."
- "The diagonals are congruent."
- "Adjacent sides are perpendicular."

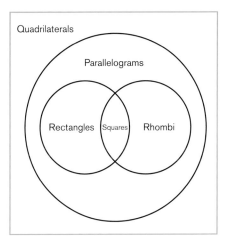

The students' brainstorming can provide an excellent opportunity to review specialized vocabulary and reexamine properties of rectangles. Some students may volunteer the fact that a square is a special kind of rectangle with all sides equal. Some may use precise terms such as *quadrilateral* and *parallelogram*. Thinking about and discussing the characteristics of a rectangle can help students move beyond basic properties as they investigate relationships in a set of rectangles.

Explore

The investigation Reasoning about Similar Figures has two activity sheets: "Diagonal Discoveries" and "Rectangles for 'Diagonal Discoveries.'" Give every student a copy of each sheet, and arrange the students in groups of two or three to work on the investigation collaboratively. Distribute scissors and rulers to each group. Direct the students' attention to step 1 on the activity sheet "Diagonal Discoveries." In this step, the students cut out the five rectangles—each with vertices labeled *A*, *B*, *C*, and *D*—which appear on "Rectangles for 'Diagonal Discoveries.'" Emphasize that the students should take care with the scissors, cutting on the lines as cleanly as possible, to preserve each rectangle's dimensions.

When the students have finished cutting, direct them to work together on the remaining steps on the sheet "Diagonal Discoveries." Urge them to discuss their ideas within their groups. Receiving feedback from peers can help them refine their ideas and supporting arguments. Your input can be particularly valuable at this time as well. Move around the classroom as your students work, adding your own comments to theirs.

In step 2, the students must enter data in a table to show each rectangle's length, width, perimeter, area, and ratio of width to length (see fig. 3.2). Stress the fact that the completed table will provide data that can help them investigate the characteristics of the rectangles.

In step 3, the students inspect the data to see if they can find a relationship among rectangles in the set. The ratios of width to length, expressed in lowest terms, are the same for three of the rectangles. However, the students may not recognize this relationship or understand its significance at this stage.

The remainder of the activity sheet guides the students in probing the relationship among the diagonals of similar rectangles. Step 4 takes the students through the process of drawing diagonal *AC* on each rectangle and "nesting" all five rectangles with all vertices *A* coinciding and all sides *AB* aligned and all sides *AD* aligned as well (see the margin).

This "stacking" will help the students answer the question in step 5: "What relationship do you find among the diagonals of some of the

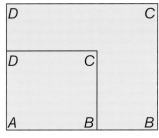

"Nested" Rectangles

Rectangle	Length (cm)	Width (cm)	Perimeter (cm)	Area (cm²)	$\dfrac{\text{Width}}{\text{Length}}$
Rectangle 1	5	4	18	20	$\dfrac{4}{5}$
Rectangle 2	8	6	28	48	$\dfrac{6}{8}$, or $\dfrac{3}{4}$
Rectangle 3	15	12	54	180	$\dfrac{12}{15}$, or $\dfrac{4}{5}$
Rectangle 4	18	14	64	252	$\dfrac{14}{18}$, or $\dfrac{7}{9}$
Rectangle 5	20	16	72	320	$\dfrac{16}{20}$, or $\dfrac{4}{5}$

Fig. **3.2.**

The students enter data on the set of five rectangles in a table.

rectangles?" Students should notice that in the case of the three rectangles whose ratios of width to length are equal, the diagonals all lie in the same line.

In step 6, the students use their work in the activity to make a generalization about this relationship. Students who are sufficiently familiar with the concept of similarity will probably conclude that these three diagonals "line up" because their respective rectangles are similar.

Step 7 asks the students to construct mathematical arguments to support the generalizations that they came up with in step 6. Give each student one or two sheets of centimeter grid paper to use in building an argument. Drawing the rectangles on a coordinate grid can help students explore their generalizations and advance their justifications.

One approach to an explanation of the relationship that the students discover depends on coordinate geometry and finding the equation of a line—specifically, the line of the diagonals of the similar rectangles. Transformations—more particularly, *dilations*—can provide another approach. You might suggest to students who have difficulty formulating support for their conjectures that they consider working with the concept of a dilation to support their conjecture.

When all the students have finished, reconvene the class and ask all the students to share their ideas. After the discussion, you may wish to give additional time for students to make changes before turning in their final products.

Evaluate

Classroom trials of the investigation provide the basis of the following discussion of students' work in the investigation. The approaches to the problem used by students in these classes may help you evaluate your students' ideas.

Step 2 of the investigation asks students to collect and display data on the five rectangles in the set. An analysis of the data (step 3) can quickly reveal that rectangles 1, 3, and 5 have a common width-to-length ratio, expressed in simplest terms as $\frac{4}{5}$. The ratio for rectangle 3 is $\frac{12}{15}$, or $3 \times \frac{4}{5}$, and the ratio for rectangle 5 is $\frac{16}{20}$, or $4 \times \frac{4}{5}$. Thus, the *scale factors* for comparing these rectangles to the smallest rectangle in the set (rectangle 1) are 3 and 4, respectively. Rectangle 2 has a width-to-length ratio of $\frac{6}{8}$, or $\frac{3}{4}$, and the ratio for rectangle 4 is $\frac{14}{18}$, or $\frac{7}{9}$. Consequently, neither of these rectangles has sides that are proportional to corresponding sides in the other rectangles.

In one of the classrooms trials, not a single student simplified the ratios to determine which rectangles were similar. Instead, most students used cross multiplication. They were comfortable in applying an algorithm to determine which ratios were proportional, although they did not demonstrate a conceptual understanding of the property.

After cross multiplying, one student applied the transitive property to conclude which ratios were proportional. Inspecting the cross products, this student concluded that $\frac{4}{5}$ was proportional to $\frac{12}{15}$, and that $\frac{12}{15}$, in turn, was proportional to $\frac{16}{20}$. Then, without additional computation, the student remarked, "so 4 : 5 must be proportional to 16 : 20…. So rectangles 1, 3, and 5 are similar" (see fig. 3.3).

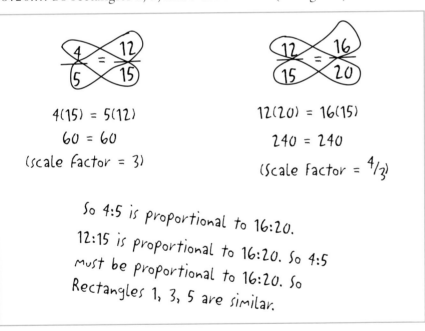

Fig. **3.3.**

A student's use of the transitive property of equality and products of cross multiplication to establish the proportionality of rectangles

It is important for students to know how similarity is defined. Similar figures have corresponding angles that are congruent and corresponding sides that are proportional—that is, they are in a constant ratio. Any discussion of similarity should also introduce or reinforce the concept of a *dilation*. A dilation is an image of a figure that enlarges or reduces the original figure. In a dilation, the image and the preimage—the original figure—are similar figures by the definition given above—they have corresponding angles that are congruent and corresponding sides that are in the same ratio. Figure 3.4 illustrates the similarity of pentagon

ABCDE and a dilation, pentagon *A'B'C'D'E'*. The dilation is an enlargement of the original figure, and the scale factor is 2.

Fig. **3.4.**

The process of dilation produces an image
that is similar to the original figure

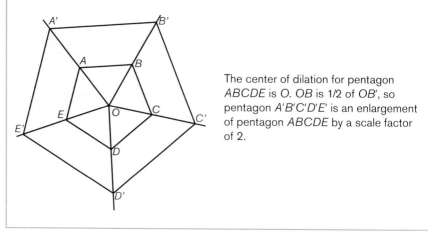

The center of dilation for pentagon *ABCDE* is *O*. *OB* is 1/2 of *OB'*, so pentagon *A'B'C'D'E'* is an enlargement of pentagon *ABCDE* by a scale factor of 2.

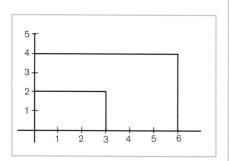

Fig. **3.5.**

Relationships of areas and scale factors
in similar rectangles

Suppose that *k* is the scale factor of an image. If *k* > 1, the image is an enlargement of the preimage. If 0 < *k* < 1, then the image is a reduction of the preimage. If *k* = 1, then the resulting image and the preimage are congruent.

Figure 3.5 also shows similar figures—in this instance, rectangles. The rectangle that is 4 units by 6 units is an enlargement of the rectangle that is 2 units by 3 units. By the same token, the rectangle that is 2 units by 3 units is a reduction of the rectangle that is 4 units by 6 units. The scale factor for comparing the two rectangles is either 2 or $\frac{1}{2}$ depending on the "direction" of the dilation. Does it enlarge, or does it reduce? The dimensions of the 4-by-6 rectangle are twice as large as those of the 2-by-3 rectangle; the dimensions of the 2-by-3 rectangle are $\frac{1}{2}$ as large as those of the 4-by-6 rectangle.

Students' work in the classroom trials revealed some misconceptions that teachers assisted in dispelling. Several students wrote in their arguments that the diagonals of a rectangle bisect the angles of the rectangle. Working with dynamic geometry software proved to be a good way to disabuse students of this notion. They were able to construct a rectangle and its diagonals and display angle measurements (see fig. 3.6). By changing the dimensions of the rectangle and examining the changing angle measurements, they discovered that this conjecture is not true for all rectangles but only in the special case of a square.

Squares are special cases of rhombi as well as of rectangles. Squares have four right angles, like all rectangles, and they have four congruent sides, like all rhombi. Thus, students might conjecture that the diagonals bisect the angles in the case of all rhombi—not just squares. Working with geometry software can provide evidence that the proposition is true in the general case for a rhombus. As in the case of a square, the diagonals of a rhombus divide it into four congruent triangles and thus bisect each of the rhombus's angles.

In working with the rectangles in the investigation, students in the classroom trials sometimes mistook diagonals for lines of symmetry. In some cases, this misconception arose from observing—correctly—that a diagonal divides a rectangle into two congruent triangles (see fig. 3.7*a*). Teachers were often successful in helping students dispel the idea that the diagonal is therefore a line of symmetry by asking the students to work with a Mira (see the margin) to determine that a diagonal does not divide a rectangle into parts that are mirror images of each other. Alter-

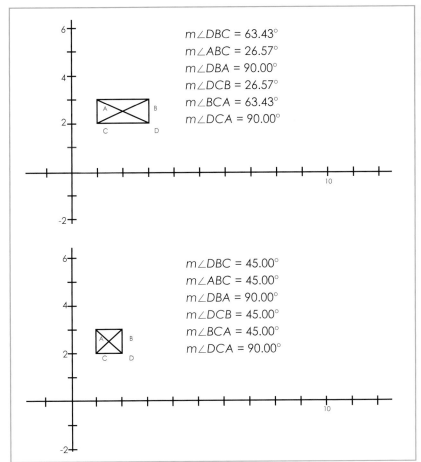

$m\angle DBC = 63.43°$
$m\angle ABC = 26.57°$
$m\angle DBA = 90.00°$
$m\angle DCB = 26.57°$
$m\angle BCA = 63.43°$
$m\angle DCA = 90.00°$

$m\angle DBC = 45.00°$
$m\angle ABC = 45.00°$
$m\angle DBA = 90.00°$
$m\angle DCB = 45.00°$
$m\angle BCA = 45.00°$
$m\angle DCA = 90.00°$

Fig. 3.6.

Working with dynamic geometry software helped students reject the conjecture that the diagonals of a rectangle bisect its angles.

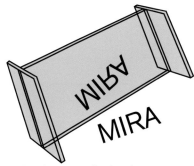

A Mira is a small plastic instrument that students can use to create a reflection of a figure and draw its image by looking through the plastic.

natively, they can fold the cut-out rectangle on a diagonal and observe that the two triangles do not align.

Students should understand that lines of symmetry reflect a figure onto itself. In their work with reflective devices, they should discover that using a diagonal as a line of reflection produces an image that is a kite—not a rectangle. Encourage your students to find a rectangle's two lines of symmetry (see fig. 3.7*b*).

After the students inspect the data in step 3 and determine that three of the rectangles are similar, step 4 asks them to draw a diagonal on each rectangle in the set of five. They then stack ("nest") all the rectangles in such a way that they can think about any relationships among the diagonals (see fig. 3.8).

In the classroom trials, some students made the true conjecture that the diagonals of the similar rectangles are in the same proportional relationship as the corresponding sides. One student who made this conjecture stated that she knew from earlier work that corresponding bases and corresponding heights of similar rectangles are proportional. She used the Pythagorean theorem to show that the diagonal of rectangle 1 (whose width is 4 centimeters and whose length is 5 centimeters) is $\sqrt{5^2 + 4^2}$, or $\sqrt{41}$, centimeters. She then showed that the diagonal of rectangle 3 (whose width is 12 centimeters and whose length is 15 centimeters) is $\sqrt{12^2 + 15^2} = \sqrt{144 + 225} = \sqrt{369}$ centimeters.

She noted that $\sqrt{369}$ is equal to $\sqrt{9 \times 41}$, which she simplified to $3 \times \sqrt{41}$. Thus, she could conclude that the diagonals of rectangles 1

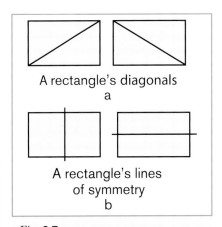

A rectangle's diagonals
a

A rectangle's lines of symmetry
b

Fig. 3.7.

A diagonal divides a rectangle into two congruent triangles that are not reflections of each other; a line of symmetry divides the rectangle into two congruent rectangles that are reflections of each other.

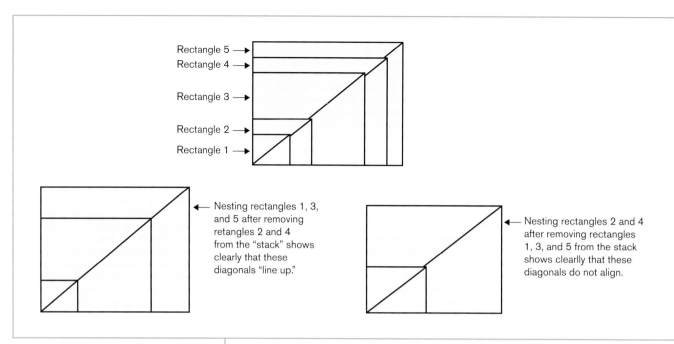

Rectangle 5 →
Rectangle 4 →
Rectangle 3 →
Rectangle 2 →
Rectangle 1 →

← Nesting rectangles 1, 3, and 5 after removing retangles 2 and 4 from the "stack" shows clearly that these diagonals "line up."

← Nesting rectangles 2 and 4 after removing rectangles 1, 3, and 5 from the stack shows clearly that these diagonals do not align.

Fig. **3.8.**

Drawing a diagonal on each rectangle and aligning ("nesting") the rectangles helps students observe relationships among the diagonals.

and 3 differ by a factor of 3, as do the sides, and the scale factor for the sides applies to the diagonals as well.

The nesting of the rectangles allows for further investigation of relationships among the diagonals of the rectangles, helping students see that the diagonals of the similar rectangles "line up," giving them a basis for conjecturing that the diagonals of similar rectangles are in the same line. Figure 3.9 shows the convergence at A of all of the diagonals that the students marked on the rectangles, giving a clearer view of the three distinct lines that these diagonals form than the representation of the stack of opaque paper rectangles in figure 3.8. Students in the trial classrooms offered such statements as, "If two or more rectangles are similar then their diagonals are collinear," and, "All diagonals for proportional rectangles match up."

These conjectures required mathematical justification. With some prompting, most students drew the rectangles in the Cartesian coordinate system to show that the diagonals of rectangles 1, 3, and 5 are in the same line. Teachers found that students often needed to be reminded to use grid paper as a tool to help them analyze the lines defined by the diagonals of the similar rectangles. Some students needed even more explicit guidance, through such questions as, "Can you show that the lines forming the diagonals represent the same line? How can you do this?" Working with a Cartesian grid in this manner can give students a good foundation for later work with right triangle trigonometry.

Some students in the trials demonstrated remarkable skill in working with the rectangles in the coordinate plane. They discovered that placing the nested rectangles on the grid with the origin (0, 0) as vertex A of each rectangle made the job of finding equations of the lines much easier. They made use of the idea of *slope*, often expressed as

$$\frac{\triangle y}{\triangle x}, \frac{y_2 - y_1}{x_2 - x_1}, \text{ or "rise over run." Once these students determined that}$$

the diagonals of the similar nested rectangles have the same slope and y-intercept, they concluded that the diagonals are, indeed, collinear.

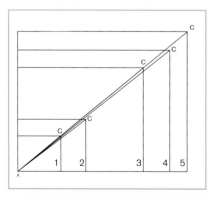

Fig. **3.9.**

The diagonals AC that the students draw on the five rectangles define three lines that intersect at A.

Figure 3.10 shows one student's work with a grid and Cartesian coordinates to show that all three diagonals have the same slope and lie in the line whose equation is $y = \frac{4}{5}x$.

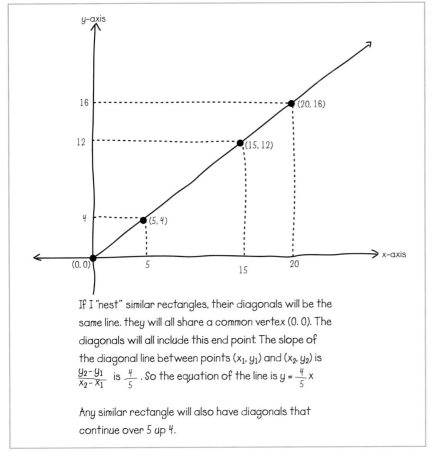

If I "nest" similar rectangles, their diagonals will be the same line. they will all share a common vertex (0, 0). The diagonals will all include this end point. The slope of the diagonal line between points (x_1, y_1) and (x_2, y_2) is $\frac{y_2 - y_1}{x_2 - x_1}$ is $\frac{4}{5}$. So the equation of the line is $y = \frac{4}{5}x$

Any similar rectangle will also have diagonals that continue over 5 up 4.

Fig. **3.10.**

A student's use of the Cartesian coordinate system to show that the diagonals of rectangles 1, 3, and 5 all lie in the line whose equation is $y = \frac{4}{5}x$

Several students noticed that the slope of the line is the same as the simplified ratio for the sides of the similar rectangles. When asked about the relationship between that ratio and the scale factor, a student exclaimed, "Ohhh, you just do over 5 up 4 either once, three times, or four times." The teacher needed to help the student understand the relationship between the equations of the lines and the concept of similarity in the rectangles.

Students in another classroom had recently worked with dilations, and many of them drew on this work as they constructed arguments to support their conjectures about the diagonals of similar rectangles. One student applied an underlying principle that for every point P in a figure, a dilation maps an image, P', on a line that passes through P and a point O—the *center of dilation*—at a distance from O that is determined by a scale factor. These students argued that aligning the rectangles so that they share a vertex provides a center of dilation not only for the sides of the rectangles but for the diagonals as well. This relationship can be expressed more formally:

The dilation of a point P through a point O by a scale factor k determines a point P' according to the following rule:

If $k > 0$, then P' lies on ray OP and $OP' = k \times OP$;

if $k < 0$, then P' lies on ray PO and $OP' = |k| \times OP$.

Figure 3.11 shows the work of a student who applied an understanding of dilations to argue that in a dilation of a rectangle *ABCD* through vertex *A*, the dilated diagonal *AC'*—the image of diagonal *AC*—lies in ray *AC*, which has a slope of $\frac{4}{5}$ and a length that is equal to the product of the length of the original diagonal and the scale factor that determines all the distances from *A*, the center of dilation, for the dilated image of rectangle *ABCD*.

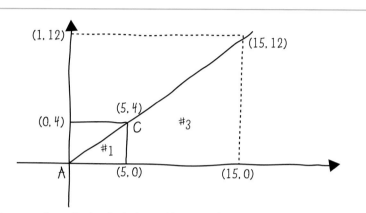

Let's see … if we think of dilations, then it makes sense that any side of a similar figure can be extended by some number (scale factor). My example rectangle 1 can be dilated by extending each side 3 times (4 X 3 = 12 and 5 X 3 = 15). The diagonal can also be "projected" 3 times to vertex of similar rectangle #3. All points from 1 rectangle will be projected the same distance – on the same line.

The materials that middle-grades teachers provide and the questions that they ask naturally influence their students' approaches to mathematical justifications for conjectures made in an investigation. Having access to grid paper induced many students in the classroom trials to represent the rectangles on a Cartesian grid. This approach appeared to facilitate their conclusion that the diagonals *AC* of the nested similar rectangles were in the same line. Working with the Cartesian coordinate system gave these students a way to reason about and demonstrate this idea successfully to themselves and others. As indicated above, other students' recent experiences with dilations promoted a different approach. Applying ideas about dilations proved to be a powerful alternative way of arguing that these diagonals lie in the same line. Both approaches provided foundations for effective arguments about the diagonals of the similar rectangles, and subsequent whole-class debriefing sessions helped students recognize that valid conclusions can be based on different mathematical approaches and principles.

Extend

Students can extend their work in the investigation in a number of ways. Perhaps the most obvious next step is to consider whether they can generalize their conjectures and arguments about diagonals of nested similar rectangles to diagonals of all similar polygons that are nested in the same way. Other natural extensions that students can explore fruitfully are the relationships between the respective perimeters and areas of similar rectangles. They can go on to consider whether the relationships that they find hold for all similar figures.

Students can discover that a scale factor of n for similar rectangles means that the perimeters of the rectangles are related by a factor of n and the areas of the rectangles are related by a factor of n^2. For example, the scale factor for the similar rectangles in figure 3.5 is 2. Can your students explain mathematically why the perimeter of the larger rectangle is twice that of the smaller one? They should be able to offer a justification like the following:

$$P_{smaller} = l + l + w + w = 2l + 2w = 2(l + w)$$

$$P_{larger} = 2l + 2l + 2w + 2w = 4l + 4w = 4(l + w) = 2(2(l + w)).$$

As a result of examining examples like this one, students should conjecture that in general the perimeters of similar rectangles are related by a factor that is equal to the scale factor.

A visual inspection of the situation in figure 3.5 might induce students to say that the area of the larger rectangle is four times that of the smaller. Can they offer a mathematical explanation in support of this idea? You might ask, "Why is it true that the area of the larger triangle is four times the area of the smaller one?"

See if your students can relate this fact to the scale factor. Because the width and the length of the smaller rectangle are both multiplied by 2 to yield the width and length, respectively, of the larger rectangle, the area of the larger rectangle is equal to that of the smaller rectangle multiplied by 4:

$$A_{smaller} = l \times w$$

$$A_{larger} = 2l \times 2w = 4(l \times w).$$

Be sure that the students understand that the factor by which the areas of similar figures are related is the square of the scale factor, since the scale factor is the multiplier for both linear dimensions in the dilation. Thus, when n is the scale factor for two similar rectangles, the factor by which the areas of the rectangles are related is n^2. By recognizing that the scale factor extends the dimensions of the original rectangle by a factor of n, the students discover that they could decompose the larger rectangle into $n \times n$, or n^2, rectangles with the original dimensions.

Students might also extend their reasoning by exploring the volumes of similar figures. They might construct an argument to explain why a cube scaled by a factor of n has a volume that is n^3 times the original volume.

Indeed, you can extend your students' experiences with geometric reasoning by offering them a variety of problems like the one in the investigation—problems that have the potential to lead gradually to formal reasoning and proof. As middle school students advance in their reasoning, they should develop an understanding of a counterexample as a tool for overturning a conjecture persuasively. The sample problems below outline a trajectory from very preliminary "warm-up" exercises to problems suited to the use of a counterexample. These problems can provide ideas about the types of experiences that students

Students can use interactive applets on the Web to dilate three-dimensional figures by a scale factor and see the effects on surface area and volume. See, for example, http://www.shodor.org/interactivate/activities/sa_volume/index.html.

in grades 6–8 need for a solid foundation for later, more formal geometric reasoning and proof.

Warm-Ups for Geometric Reasoning

Early experiences in geometric reasoning should build on very basic information that students know and engage them in extending that information by creating convincing arguments about another geometric principle with which they are less familiar. Consider the following problem:

> The sum of the angle measures of a triangle is 180 degrees. Construct an argument to convince someone that the sum of the angle measures of a quadrilateral is 360 degrees.

This problem may appear intuitive, but the goal is to "scaffold" both learning about geometry and skill in constructing a mathematical argument, enabling students to move with maximum independence from known facts to ideas that are new to them. Students must use geometric reasoning to construct an argument and solve the problem, but their argument need not constitute a formal proof. Where should they begin?

They can decompose a quadrilateral into two triangles (see fig 3.12a). This process, they discover, involves using adjacent sides of the quadrilateral as two sides of a triangle and forming the triangle's remaining side by drawing a diagonal of the quadrilateral. How many segments can they draw that decompose a rectangle into two triangles? The students can argue that either diagonal of an arbitrary quadrilateral decomposes the figure into exactly two triangles. Because they know that the sum of the angle measures in a triangle is 180 degrees, they can conclude that sum of the angles in the two triangles is 2×180, or 360, degrees.

After students reach this point, invite them to consider any irregular convex n-gon that you draw on the board (see the samples in fig. 3.12). Select one vertex, and draw all the diagonals that connect it to other vertices. Prompt your students to observe that the selected vertex is connected to the two adjacent vertices by sides of the n-gon, and that the vertex cannot be connected to itself. Therefore, you can draw $n - 3$ diagonals from the selected vertex. From their explorations with quadrilaterals, students should conclude that the procedure decomposes any n-gon into $(n - 3) + 1$, or $n - 2$, triangles.

Reasoning by Counterexample

Asking, "Can you think of a counterexample?" can motivate students to consider multiple examples related to a particular phenomenon or property, giving them an opportunity to find one example that reveals that a conjecture does not hold in all cases. Using a counterexample is a basic and indispensable way of showing that a conjecture is not always true. An argument that supplies just one counterexample demonstrates that a given proposition is false. The following problem can serve to introduce middle school students to the power of a counterexample:

> Two student are discussing triangles:
> *Adele:* "Triangles that have the same area are congruent."
> *Yan:* "That can't be true. I'll show you an example to prove that it's not."

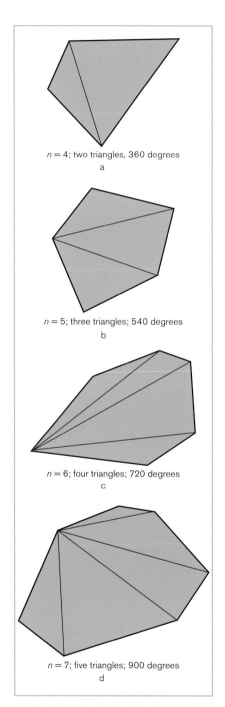

$n = 4$; two triangles, 360 degrees
a

$n = 5$; three triangles; 540 degrees
b

$n = 6$; four triangles; 720 degrees
c

$n = 7$; five triangles; 900 degrees
d

Fig. 3.12.

Decomposing n-gons into triangles

Navigating through Problem Solving and Reasoning in Grades 6–8

Can you suggest an example that Yan could use to convince
Adele that her statement is false?

Many students would be likely to approach this problem through
trial and error, testing examples until they found one that didn't work.
Urge your students instead to work systematically, thinking about
relevant concepts. Encourage them to consider what variables have an
impact on the situation. Given that the area of a triangle is equal to $\frac{1}{2}$
(bh) with b as the triangle's base and h as its height, the students should
see that focusing on the base or height can be an effective way to
approach the problem.

Students probably know that a general definition of congruent figures includes two conditions. For figures to be congruent, (1) all corresponding sides must be congruent, and (2) all corresponding angles
must also be congruent. As a result of these conditions, congruent figures have the same area as well as the same shape. Thus, if the students
find just one pair of triangles whose areas are the same but whose shapes
are different, they will have shown that Adele's conjecture is false.

Although one such example will suffice, you might encourage your
students to find several counterexamples to Adele's idea. The discovery
of these counterexamples can reinforce their recognition of the usefulness of a counterexample as well as their understanding of the important geometric idea that polygons of equal area need not be congruent.
The counterexamples in figure 3.13 reflect three approaches to the
construction of a counterexample to Adele's proposition—and illustrate
the advantages of working on a grid in such a case.

"More than any
other area of
mathematics, the
geometry
curriculum abounds with
opportunities for students to
investigate if a conjecture is
always, sometimes, or never
true." (Boats et al. 2003, p. 210;
available on the CD-ROM).

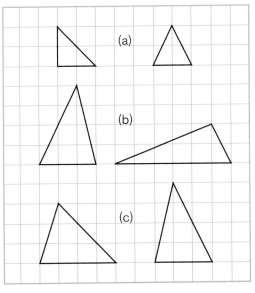

Fig. **3.13.**

Three counterexamples to the proposition
"All triangles of equal area are congruent"

Figure 3.13*a* shows a counterexample constructed by drawing two
triangles that have congruent bases and congruent heights but are
nonetheless obviously not congruent. Despite having equal areas of
$\frac{2 \times 2}{2}$, or 2, square units, one of the triangles is a right triangle,
and the other is acute.

The counterexample in figure 3.13*b* results from a different approach—drawing triangles whose bases are not congruent and whose

heights are also not congruent but whose base and height yield the same product in both cases: 3 × 4 = 12; 6 × 2 = 12. Again, despite having equal areas—this time of 6 square units—the triangles do not have corresponding sides and corresponding angles that are congruent. Hence, the triangles are not congruent.

The counterexample in figure 3.13c reflects a third strategy—drawing a triangle and then drawing another that simply reverses the base and height of the first. Of course, the areas of the triangles in this case are the same: $\frac{3 \times 4}{2}$, or 6, square units. But their corresponding sides and corresponding angles are not congruent, so the triangles are not congruent.

An alternative argument about Adele's conjecture might begin with a square and a rectangle that is not a square but whose area is equal to that of the square. Students might, for example, take a rectangle with a length of 9 and a width of 4 and a square with a side of 6 units. Both shapes have areas of 36 square units (see fig. 3.14). However, the square and nonsquare rectangle are not congruent—their respective bases and heights are both different. By drawing one diagonal of each shape, the students can divide both into two congruent triangles. If they take one triangle from the square and one from the nonsquare rectangle, they have two triangles, each with an area of 18 square units—half of the area of the shapes from which they came. But the bases of these triangles are different, as are their heights, so the triangles have different shapes and are not congruent. This single counterexample is sufficient to refute Adele's conjecture.

Fig. **3.14.**

A nonsquare rectangle and a square of equal area, each divided by a diagonal, producing four triangles of equal area with neither triangle in the nonsquare rectangle congruent with either triangle in the square

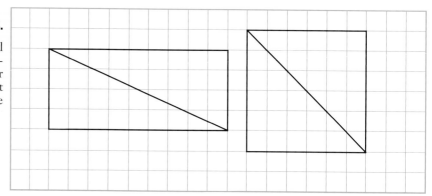

Another possible argument about Adele's idea begins with a rhombus—a quadrilateral with all sides congruent. Students might draw two congruent rhombi (see fig. 3.15). They can divide each by a diagonal into the pairs of congruent triangles shown. The area of each triangle is 8 square units—half of the area of each rhombus. Each diagonal forms a side of the two mirror-image triangles that it creates. However, the diagonals are not congruent—one is 4 units, and the other is 8 units. So a triangle that composes half the area of one rhombus is not congruent to a triangle that composes half the area of the other, because these triangles have only two congruent sides—not three, as required for congruence. Again, the single counterexample has the power to overturn Adele's conjecture.

When middle school students find counterexamples and construct arguments based on them, they engage in powerful reasoning about principles and properties of geometry. They gain valuable experience in looking for patterns, defining relationships, and considering cases and examples as means of determining the validity of statements. These habits of mind prepare them for the more rigorous, formal thinking that they will need for success in high school.

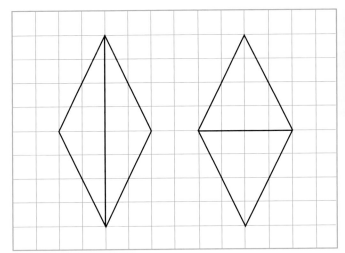

Fig. **3.15.**

Two congruent rhombi, each divided by a diagonal, producing four triangles of equal area with no triangle in either rhombus congruent with a triangle in the other

Conclusion

Geometric reasoning is a complex skill that develops gradually. Meaningful and structured explorations like those presented in this chapter promote its growth. As illustrated in the investigation Reasoning about Similar Figures, such experiences call for inquiry, conjecturing, and constructing arguments about geometric ideas, leading gradually to formal proofs in geometry. Middle school mathematics classrooms can be rich environments for the development of geometric reasoning, a strand of mathematical reasoning that is crucial to students' success in higher-level mathematics and their effectiveness as flexible problem solvers. The next chapter illustrates the potential for the development and use of reasoning in grades 6–8 in another essential branch of mathematics—data analysis.

PROBLEM SOLVING *and* REASONING

Chapter 4

Reasoning about Data Analysis

Statistics is a relatively new but vital topic in the K–12 mathematics curriculum (see NCTM [1989, 2000]). Before the 1990s, most courses in statistics and data analysis were offered at the postsecondary level. David Moore (1991) makes a compelling argument for the teaching of statistics in school mathematics: "If the purpose of education is to develop broad intellectual skills, statistics merits an essential place in teaching and learning" (p. 134). Moore argues that statistical thinking, broadly defined, "should be part of the mental equipment of every educated person" (pp. 134–35). David Ben-Zvi and Joan Garfield (2004) characterize statistical reasoning "as the way people reason with statistical ideas and make sense of statistical information" (p. 7). They elaborate on what such reasoning entails:

> This involves making interpretations based on sets of data, representations of data, or statistical summaries of data. Statistical reasoning may involve connecting one concept to another (e.g., center and spread), or it may combine ideas about data and chance. Reasoning means understanding and being able to explain statistical processes and being able to fully interpret statistical results. (Ben-Zvi and Garfield 2004, p. 7)

The learning of statistics includes two central components: a mastery of the concepts and conventions of the subject and an ability to apply this knowledge in a broad, flexible, and skeptical manner to analyze data and interpret or communicate results. This knowledge and skill nurture a facility with statistics—a "statistical gestalt"—that is both difficult to describe and challenging to develop. Advances in the teaching of statistics in K–12 classrooms depend on the growth of knowledge about ways to support students' development of statistical concepts and statistical reasoning. Despite the challenges of teaching statistics, educators usually agree that statistics is a practical subject, that students need to become competent in carrying out statistical investigations, and that in general successful members of society need to be "intelligent consumer[s] of data" (Shaeffer, Watkins, and Landwehr 1998, p. 4).

Engaging Students in Statistical Reasoning

Research in statistics education (Bakker 2004) indicates that students often learn statistics as a set of techniques that they apply indiscriminately. Many of the tasks that teachers offer in the elementary and middle grades make low-level cognitive demands on students instead of giving them opportunities to use higher-level thinking to make sense of the situations under investigation. Teachers and curriculum specialists need to design tasks that are sufficiently demanding to facilitate an appropriate cognitive development in the students' understanding of the concepts and conventions of statistics. Such tasks must involve contexts that call for genuine statistical reasoning.

Newspaper articles, televised reports, stories in consumer magazines, and accounts in other media frequently present analyses of data. The results of such analyses are useful to many people in making decisions and developing arguments. Students need to learn how to read and interpret the results of data analyses and be prepared to reason about, and on the basis of, data.

Statistical reasoning is different from knowing the procedures for carrying out different kinds of analyses, such as finding a mean or making a graph. Reasoning from and about data occurs in a fluent, meaningful way when someone formulates a question, determines what information to collect, gathers appropriate data, decides how to analyze them, and performs the analysis to yield results that answer the question. Making decisions based on data often involves sifting through a great deal of information.

Clarity is essential to such reasoning. A clearly formulated question, fully understood, guides decisions about what information to gather to answer the question. An unclouded understanding of the context is a prerequisite to the formulation of the question, which comes out of and is closely tied to the context. A detailed understanding of the context also guides the collection of data and the mathematics used to analyze them. A sure grasp of the connection between the context and the mathematics is indispensable as well, since the mathematics must suit the context—in some way modeling it—to give results that are meaningful and useful.

Statistical reasoning in middle school—a case study

Middle school students need practice in working with data to learn the importance of examining a context as well as their own perspectives on it and any assumptions that they make about it. They should experience firsthand the need for clear connections among context, question,

Navigating through Data Analysis in Grades 6–8 (Bright et al. 2003) offers a variety of problem contexts for the application of statistical reasoning.

data, assumptions, and analysis as they solve problems by working with data.

Whose tour was most successful?

Consider, for example, the work of middle school students in determining which of four music industry headliners—Barbra Streisand, Boyz II Men, the Eagles, or the Rolling Stones—had the most successful tour in a particular season (see Mooney [2002]). The students came up with a variety of measures, such as how much money a tour made and how many performances it included. Drawing data from the *World Almanac and Book of Facts* (Famighetti 2000), they made four different bar graphs that displayed total concert earnings, numbers of performances, numbers of cities visited, and average earnings per show (see figs. 4.1–4.4; data from Famighetti [2000, p. 186]).

The students then examined each graph to decide how to interpret the information to help them answer the question about which headliner's concert tour was the most successful. They also pondered

The bar graphs that display data on the tours by Barbra Streisand, Boyz II Men, the Eagles, and the Rolling Stones also appear on the CD-ROM as a set of blackline masters titled "Whose Tour Was Most Successful?" for use with students in a data analysis exercise.

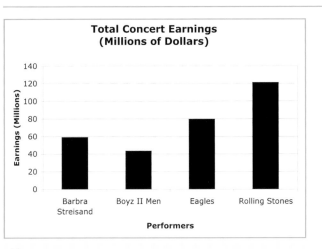

Fig. **4.1.**

Four headliners' total earnings from their concert tours n one season

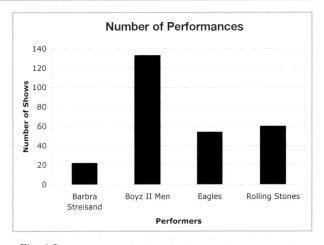

Fig. **4.2.**

Four headliners' numbers of performances in their concert tours in one season

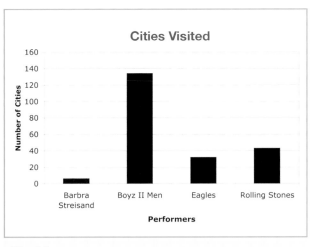

Fig. **4.3**.

Numbers of cities visited by four headliners in their concert tours in one season

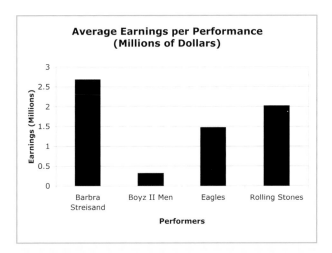

Fig. **4.4.**

Four headliners' average earnings per performance in their concert tours in one season

questions that the data did not give them information to answer. For example, they wondered about the data on average earnings per performance. The students recognized that the statistics reported in the graph in figure 4.4 depended on the number of shows that headliners had performed, but some students wondered whether each headliner had given the same number of shows, or whether a headliner with smaller earnings per show had actually given many more performances than some other headliner with greater earnings per show. Others wondered whether the particular cities in which the headliners had performed mattered? Did the headliners draw larger crowds in some cities than in others? The students learned that these questions were not answerable on the basis of the data. They also discovered that they could not resolve some of the questions raised by their inspection of one graph by inspecting data in another graph.

Figure 4.5 shows responses of four students to the question, "Whose tour was most successful?" The answers illustrate different interpretations of success and the measures that determine it.

Fig. 4.5.

Four students' analyses of the data in response to the question, "Whose tour was most successful?" (adapted from Mooney [2002])

Scott:	I added up the heights of the bars of Boyz II Men, and that's higher than Barbra Streisand and higher than the other two groups.
Nancy:	Well, … on each graph I noticed who had … the highest amount, and two out of four times Boyz II Men had the highest, and Barbra Streisand only had the highest once, and Rolling Stones only had it once, [and the Eagles never were highest].
Charlene:	Well … Probably the Rolling Stones. Yeah. They had the highest total concert earnings.
Jayson:	If you gauged "successful" on the money made, then probably the Rolling Stones were the most successful because they made the most money. But if you say success is based on the [average] earnings per performance, then Barbra Streisand was probably the most successful. But if … you took these bars and put them one on top of the other, and you stacked them all up, I guess whichever one would be the tallest would be the most successful.

Scott judged that Boyz II Men had the most successful tour. He mentally stacked each headliner's bars from all four bar graphs and called the headliner with the tallest composite bar the most successful. In other words, he considered that each bar's height defined a certain degree of success and that the combined heights of all the bars defined the overall success of a headliner.

Nancy also decided that Boyz II Men had the most successful tour, although her analysis was different. She determined that this group had the most "highest values" (two) of any group in all the graphs.

Charlene concluded that the Rolling Stones had the most successful tour since this group made the most money, as measured by total concert earnings. (In average earnings per performance, the Rolling Stones were second to Barbra Streisand.)

Jayson took three perspectives into account, each of which led him to a different conclusion. He noted that defining success by the total amount of money earned (as Charlene defined it) would make the Rolling Stones' tour the most successful. However, defining success in terms of money earned per performance would make Barbra Streisand's tour the most successful. Then again, stacking the bars on top of each other for each headliner (the strategy that Scott adopted) would indicate that the tour by Boyz II Men was the most successful.

Each of the four students answered the same question on the basis of the same data, displayed in the same set of graphs. Yet, each reasoned differently. Neither Scott nor Nancy appeared to address the tour context in any significant way. Moreover, they did not seem to examine the assumptions behind the methods that they used to determine the most successful tour. Instead, they approached the data mechanically, considering what the bars stood for in only a cursory or superficial way, if at all.

Scott appeared to stack the bars for each headliner from all the graphs without considering what the data on each graph meant. What meaning would a bar have that combines the heights of bars representing average earnings per show, number of shows performed, total concert earnings, and numbers of cites in which concerts were performed? Scott did not appear to reflect on this question.

In her analysis, Nancy treated all the graphs, each with its own distinct meaning, as separate entities, but she then awarded the claim of "most successful tour" more or less automatically to the headliner who had the most "highest bars" on all four displays. She did not explicitly state that she perceived all four characteristics as equally important in determining a successful concert tour. But did she in fact consider no characteristic to be more meaningful than any other in determining success, as her method suggests? Moreover, she gave no clue about what she would have done if the data had put each of the four headliners highest in one of the four data displays. Nor did she suggest how she would have proceeded if Boyz II Men and the Rolling Stones both had "highest bars" in two out of the four data displays.

By contrast, both Charlene and Jayson stated or implied the characteristics that they viewed as important to a successful concert tour. Charlene used only one characteristic—total concert earnings—to determine the most successful tour. Her focus on this single characteristic implies that she considered "most money earned" to be the criterion of a tour's success. This analysis demonstrates that she was thinking concretely about the context of the data and the question.

Jayson recognized the possibility of multiple perspectives on, and reasonable assumptions about, the factors that make a tour successful in a

Students need to be aware of assumptions that they make in reasoning about data, and they need to recognize that different perspectives are possible. They also need to understand and make use of the context that gives rise to the question that they are investigating.

more explicit way than anyone else. He acknowledged that success could be measured by money earned in two different ways—either by total tour earnings or by earnings per performance. He suggested a third possibility, as well—success could also be measured by the highest stack resulting from combining the bars for each headliner on all four graphs.

Important considerations in reasoning about data

The different perspectives and assumptions that Jayson presented in his response to the data about the concert tours point to two important issues in reasoning about data:

1. Students need to be aware of assumptions that they make in reasoning about data, and they need to recognize that different perspectives are possible. Others can more clearly understand and more easily interpret arguments based on data if they know the assumptions on which the arguments rest and the perspectives that they reflect. Jayson asserted that the tour by the Rolling Stones was the most successful from the standpoint of total concert earnings, and the tour by Barbra Streisand was the most successful from the standpoint of average earnings per performance. Thus, he took into account two different and defensible perspectives that analyses might take in addressing the question, "Whose tour was most successful?"

2. Students need to understand and make use of the context that gives rise to the question that prompts them to gather data for analysis in the first place. Jayson offered different perspectives on the most successful tour. In his third perspective, he made use of the "stacking" method that Scott implemented. This perspective shows a lack of understanding of, or connection to, the context of the data. "Stacking" the bars from the four data displays makes no sense mathematically or contextually.

In all, Jayson provided two sound perspectives and one arbitrary and idiosyncratic one for determining a successful concert tour. He showed in his work that the context of the data was accessible to him, but he did not consistently make use of it in analyzing the data.

However, students who have little experience with data need to be aware of a different pitfall when they *do* consider the context of the data. They must avoid a tendency to bring in information that is irrelevant to the question at hand. For example, in the middle school class that was considering the tour data, one student argued that Barbra Streisand's tour was the most successful because a solo artist would not have to share the money, but ensemble artists would need to split it. The distribution of money earned in a tour is beyond the scope of the question under investigation. When students are reasoning about and on the basis of data, they need to focus on the context and the data in relation to the specific question that they are attempting to answer.

This chapter presents two investigations that call on students to reason about data. In the first, they analyze data about celebrities and establish criteria that allow them to make a case for considering one of them the most popular. In the second, they explore the use of the mean and the median—statistical measures of center—to help them understand the distribution of data in a set.

Reasoning about Data to Make Decisions

Goals

- Consider and implement various methods of making a decision or an argument on the basis of data in a table
- Identify assumptions or perspectives that shape a decision or an argument based on data
- Select and use data appropriately to make and defend a decision or argument

Materials and Equipment

For each student—

- A copy of the following activity sheets:
  - "Popular People"
  - "Who Is Most Popular?"
  - (Optional) "How Do the Dollars Stack Up?"
  - (Optional) "Word Spreads"

For each group of two to four students—

- One or two sheets of centimeter grid paper or access to graphing or spreadsheet technology

pp. 159, 160, 161–62, 163–64

The template "Centimeter Grid Paper" on the CD-ROM allows you to print out sheets of grid paper for your students' use in the investigation.

Prior Knowledge

Students should have some previous experience in creating and interpreting data in tables and graphs, including bar graphs, dot plots, and line plots.

Learning Environment

The students work in groups of two to four to analyze data in a table and answer a question. The teacher acts as a facilitator, listening carefully and asking questions to help the students articulate assumptions, perform analyses, and make decisions. When the students complete the activity, they share their decisions and explanations as a class, with the teacher prompting them to revisit ideas that lack sound reasoning or need additional support.

Discussion

In middle school, students are particularly—sometimes painfully—interested in what it means and takes to be popular. This investigation channels that energy into thinking about some of the quantifiable measures of celebrities' popularity and various ways of using data on twenty celebrities to make a defensible decision in answer to the question, "Who is most popular?"

Engage

Begin by asking your students what it means to be popular. How can they tell that someone is popular? If two people are popular, are there ways of determining that one is more popular than the other? How about in the case of celebrities? If someone is a celebrity, is he or she by definition popular? What factors would your students consider in determining the degree of a celebrity's popularity? What factors would they consider if they were ranking a group of celebrities in order of their popularity?

Let your students brainstorm to make a list of their favorite celebrities. On the basis of the names on their list, discuss what makes a celebrity popular. Use the discussion to focus the students' thinking on the different characteristics, or criteria, that someone might use to define popularity for a celebrity. Tell the students that they are going to investigate data to decide who is the most popular celebrity in a group of twenty celebrities.

Explore

Assign your students to groups of two to four, and give each student copies of the activity sheets "Popular People" and "Who Is Most Popular?" A large table on "Popular People" gives them raw data on the twenty celebrities. "Who Is Most Popular" asks them to analyze these data to justify their decisions about which three celebrities they would consider to be most popular. Allow your students time to become familiar with the data, perform their analyses, and interpret the results to draw their conclusions. The next section presents several methods used by students in classes that worked with the investigation. Their ideas illustrate some of the ways in which your students might handle the data.

Evaluate

In engaging with the data and the context, students should reflect on the various criteria that the table on "Popular People" uses as measures of a celebrity's popularity. Examining a table, as opposed to a graph, focuses students' attention on the data themselves and not, say, on the heights of bars in bar graphs. Furthermore, data are not always available in the tidy format of a graph. Sifting through data is often necessary. The task of inspecting data in the table gives the students a taste of that experience. However, the biggest challenge for them is explain how and why they reason and make particular decisions.

Many students are likely to use a rank-ordering process to determine the three most popular celebrities. For example, one group of students used a repeated elimination process. They considered four criteria—number of Web site hits, number of TV or radio reports, number of newspaper articles in which the celebrities' names appeared, and earnings. First, they choose the top ten celebrities by looking solely at the numbers of Web site hits. They then narrowed this group to a smaller one on the basis of the numbers of TV or radio reports. They reduced their new group by ordering the members according to the numbers of newspaper articles, and finally they winnowed this group to three on the basis of the members' earnings. They identified these three celebrities as the most popular.

Another group of students determined the top celebrity in each category. Because some celebrities were number 1 in more than one category, this process left the students with exactly three celebrities, whom they then regarded as the three most popular celebrities. Another group's procedure was a variation on this one. These students determined the top four celebrities in each category. Then they determined which of these celebrities appeared most often in the lists of the top four in each category and drew the top three celebrities from this new list.

Extend

It is worth noting that none of the groups whose work is discussed here considered making any graphical displays as part of their analyses of the data. The students may not have regarded these kinds of displays as useful data analysis tools (or as useful tools in this case), or they may not have had the materials or equipment—grid paper or graphing technology—to encourage them to go in this direction.

However, making or inspecting graphs can be fruitful ways for students to extend their work with data in the investigation. For example, figure 4.6 shows two dot plots that can facilitate further investigation of the celebrities' earnings. A dot plot can be a very useful tool for visualizing the distribution of data. The figure shows two measures of center—the mean and the median—for each plot.

In figure 4.6, the graphs have background shading that differentiates the graph into parts – earnings less that $50 million, and so on.

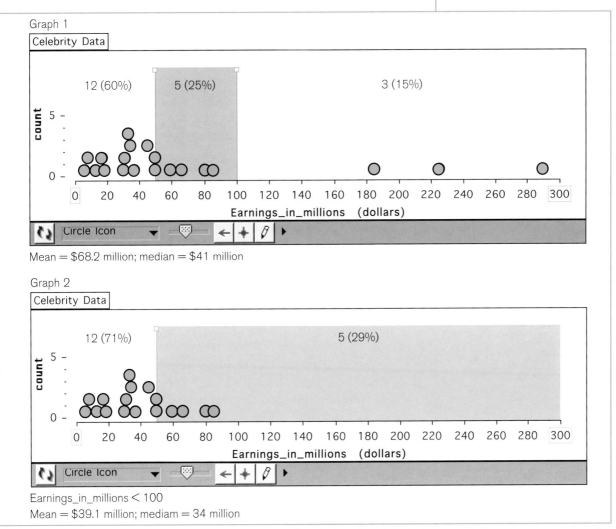

Fig. **4.6.**

Dot plots for further investigation of the celebrities' earnings

The technology used to create the plots allows users to manipulate the data in different ways for different purposes. Graph 1 illustrates the capacity of the graphing software to differentiate the data into parts to assist users in seeing where the data cluster and what points are outliers. This graph highlights the fact that the earnings of twelve celebrities (60%) were less than $50 million and the earnings of five celebrities (25%) were between $50 million and $100 million. The earnings of three celebrities (15%) were much higher—"way out on the tail" of the curve of the distribution. As in graph 1, users can easily shade regions in graphs created with technology to show such groupings of data.

Graph 2 illustrates another advantage of graphing technology. This graph shows the result of altering the first graph in such a way as to "filter out," or hide, the outliers—the earnings that were more than $100 million ($185 million for Mel Gibson, $225 million for Oprah Winfrey, and $290 million for George Lucas). Although statisticians cannot legitimately remove data from a set, they can temporarily filter out certain values to see how the distribution looks without these values. Eliminating outliers from consideration and exploring the effect on the mean, median, and other measures can be very useful in data analysis. The shading in graph 2 differentiates the data into two parts—the cluster of earnings less than $50 million and more dispersed earnings at higher figures.

The activity sheet "How Do the Dollars Stack Up?" presents the two plots from figure 4.6 for your students to work with if you decide to extend the investigation in this way. Decide how familiar the students are with the statistical meanings of the terms *variability*, *distribution*, and *measure of center*, as well as the concepts *mean* and *median*, before giving them this task.

Your students can also extend their experience with the celebrity data by inspecting other types of graphs. Figure 4.7 shows a graph that represents the data on publicity garnered in print by the celebrities, measured in numbers of magazine cover stories and newspaper articles. The graph in this figure is a scatterplot, with each data point representing these two pieces of information on a particular celebrity. Be sure that your students understand the data in figure 4.7 correctly and interpret them meaningfully. For example, call their attention to the fact that the tallies of newspaper articles include all the articles in which the celebrities' names appeared during the time period.

Some students might consider that counting every such article sets the bar very low for newspaper publicity as a measure of popularity. Bill Clinton, for instance, as a former president of the United States, is likely to be mentioned in many newspaper articles that do not focus on him. In addition, when your students consider the numbers of magazine cover stories for the celebrities, make sure that they understand what decimal fractions mean in the data. A note below the table on the activity sheet explains that a fraction of a cover story reflects a "sharing" of the story with others—for example, 0.5 represents a sharing with one other person, 0.25 represents a sharing with three others, and so on.

The activity sheet "Word Spreads" presents the scatterplot in figure 4.7 and poses questions to the students about the distribution of these data. The students must also consider what data points are outliers and

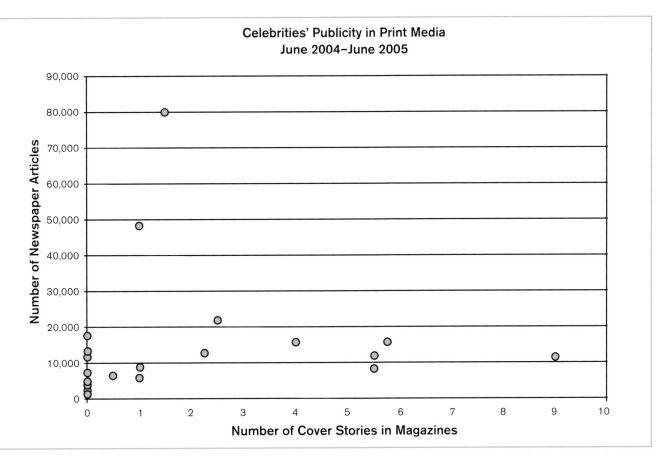

Celebrities' Publicity in Print Media
June 2004–June 2005

Fig. **4.7.**

A scatterplot showing the publicity received by twenty celebrities in print media, with numbers of newspaper articles plotted against numbers of magazine cover stories, for the period from June 2004 to June 2005

whether "filtering out" these points would be helpful to their analysis of the celebrities' popularity in terms of their publicity in print.

You can help your students extend their work with the data on celebrities in other ways as well. You might ask them to discuss the categories of data in the table in "Popular People." What do they think of these categories as criteria for determining a celebrity's popularity? Do the students need to consider data in all these categories to decide how popular one celebrity is compared with another? Can the students think of other important criteria that are missing?

Some of the students who completed the investigation did not know who George Lucas was. They wondered if "name recognition" might be an essential factor to evaluate in determining a celebrity's popularity. They believed that a lack of name recognition would make a celebrity less popular overall than other strong indicators of his or her popularity might suggest. Your students can discuss how they would measure name recognition and any other criteria that they come up with, as well as how they would use data in these new categories to determine celebrities' popularity.

Your students can also repeat the investigation in other contexts. For example, they can use the Web to collect data on baseball players or other sports figures and analyze them to determine the most successful player. Or they can update the list of celebrities for the original

For up-to-date data on celebrities' popularity, see the *Forbes* Web site. Go to www.forbes.com/lists, and under "People," locate "Most Powerful Celebrities."

investigation and reevaluate the three most popular celebrities. Recent data on celebrities' popularity are available at *Forbes* magazine's Web site.

The next investigation, Reasoning about the Mean and Median of a Distribution, reinforces important concepts of data analysis in the familiar context of watching TV. Behind this investigation lies a perennial question that adults pose to teenagers, "Are you watching too much TV?" The investigation builds on and extends activities in an earlier Navigations volume, *Navigating through Data Analysis in Grades 6–8* (Bright et al. 2003).

Reasoning about the Mean and Median of a Distribution

Goals

- Work in a concrete context to develop an understanding of core concepts in statistics, such as *mean* and *median*, as a foundation for encountering increasingly rich problems involving the use of data

- Recognize the importance of focusing on the characteristics of the distribution of data displayed in a graph to interpret data to answer a question arising in a real-world context

- Gain an increased awareness of the relationship of the mean and median—two measures of center—to the shape of the distribution of the data in a set

Materials and Equipment

For each student—

- A copy of each of the following blackline masters for the activities TV Watching and Making the Data (Bright et al. 2003, pp. 85–86; available on the CD-ROM)

  ◦ "TV Watching"

  ◦ "Making the Data"

- A copy of the activity sheet "Change the Data—Change the Stats?"

pp. 165–67

For each group of two to four students—

- Access to the applet Plop It! (available on the CD-ROM or at http://www.shodor.org/interactivate/activities/PlopIt/) or a sheet of chart paper and a pad of small sticky notes

For the teacher—

- The following materials for the activities TV Watching and Making the Data (Bright et al. 2003, pp. 23–28; 85–86; 110—all available on the CD-ROM)

  ◦ Discussions for teachers

  ◦ Blackline masters "TV Watching" and "Making the Data"

  ◦ Solutions

Prior Knowledge

Students should have explored measures of center in a number of contexts and have a basic understanding of the information that the mean and the median communicate about a distribution. Tasks in *Navigating through Data Analysis in Grades 6–8* (Bright et al. 2003) can serve as a preassessment or review of these concepts as well as an introduction to ideas about graphing and analyzing a distribution of data.

Learning Environment

The students work individually or with partners to complete two activities from *Navigating through Data Analysis in Grades 6–8* (Bright et al. 2003). Then they work in groups of two to four to enter data from one of those activities into the applet Plop It! (or on a graph on chart paper). They display the distribution and change it in various ways to analyze and interpret the impact on the *mean* and *median*. During these activities, the teacher acts as a facilitator, listening carefully and asking questions to help the students articulate assumptions, make and verify predictions, and explain results. After completing the exploration, the students come together as a class to discuss their discoveries and compare their thinking.

Discussion

Early work with data in specific contexts should develop students' understanding of such core statistical concepts as *mean*, *median*, and *mode*—concepts that enable the students to explore and analyze increasingly rich and complex problems involving the use of data. Experience in varied contexts can offer students opportunities to understand the interrelationships of these concepts and other essential statistical ideas, such as *variability* and *distribution*.

Beginning students often direct their attention to individual cases in data sets—particularly if the data provide information on themselves. With more training, however, students learn to look at the overall distribution of a data set. They recognize the importance of identifying, and in some way applying, the *characteristics* of the distribution of the data.

Any distribution has a set of statistical attributes—its mean, median, mode, range, and so on. These attributes are called the *summary statistics* of the distribution. Students should understand that when they engage in such tasks as computing the mean in baseball players' batting averages or identifying the number of boxes of raisins with exactly 35 raisins to a box in a crate of raisins, they are identifying characteristics of the distribution of the data in a set.

This chapter has illustrated several types of graphs used to show distributions of data for different purposes. Various figures have displayed bar graphs, dot plots, and scatterplots. Learning how to make different types of graphs, like learning how to obtain the mean, median, mode, and other summary information about a data set, is essential to data analysis. Moreover, this learning is more meaningful and productive when it takes place in the process of actually "doing statistics" in multiple and varied real-world contexts.

Ideally, learning how to represent data in a graph and apply concepts like mean and median should be a holistic process. Students should not only be able to compute the mean or the median but also be able to estimate their locations in a graph as they examine the distribution of data from a particular, concrete context. The students should understand the relationships among the distribution of the data, the locations of the measures of center in a graph, and the computations of their values.

Navigating through Data Analysis in Grades 6–8 (Bright et al. 2003) presents two interrelated activities back to back: TV Watching and Making the Data. These activities provide examples of the type of holistic instruction that promotes reasoning in data analysis. TV watching calls on students to—

- Make an appropriate graph for discrete data
- Describe the characteristics (e.g., clustering of data) of the graph
- Use the shape of the graph to draw conclusions about the data. (p. 23)

The sequel, Making the Data, takes the students another step forward in their thinking about data. Working in the same context—hours of TV viewing—the students suggest possible data sets that would yield the same mean or median, although their distributions might differ. Both activities are included in their entirety (discussions for teachers, blackline masters for students, and solutions) on the CD-ROM that accompanies this book. The current investigation, Reasoning about the Mean and Median of a Distribution, builds on these earlier Navigations activities, extending students' skill in reasoning about the mean and the median of a data set through the use of the applet Plop It!—available on the CD-ROM and the Web (http://www.shodor.org/interactivate/activities/PlopIt/).

Plop It! is an applet that allows users to enter data and see them displayed in a bar graph with the mean, median, and mode indicated. The applet appears on the CD-ROM and is also available on the Web at http://www.shodor.org/interactivate/activities/PlopIt/.

Engage

Guide your students in completing the activities TV Watching and Making the Data from *Navigating through Data Analysis in Grades 6–8* (Bright et al. 2003). The discussions of the activities include suggestions for introducing and facilitating them in the classroom. The students' work in the first activity focuses on making and interpreting a graph—most likely a bar graph or line plot—which shows the data in a given set of numbers of hours in a week that students in two eighth-grade classes spent watching TV. Figure 4.8 shows a bar graph of the data.

Fig. **4.8.**

A bar graph showing hours of TV watching in one week by students in two eighth-grade classes

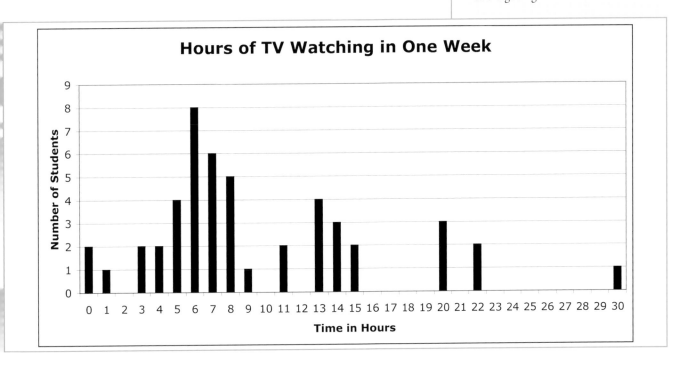

After making the graph, the students identify characteristics of the distribution of the data. They consider the number of hours per day that the majority of the students spent watching TV. They identify the greatest and smallest numbers of hours of TV watching. On the basis of their analysis of the distribution of the data, they decide whether or not they think that these students spent too much time watching TV, and they explain how they reached their conclusions.

In the second activity, the students work the other way around—from summary information about a set of data back to possible values for the data in the set. The students suppose that the members of a family of eight spend an average of five hours a week watching TV. With five hours as the mean of the data set, what might the eight values in the set be? Is more than one set of eight values possible? What if the mean is five hours and the median is four hours? Is more than one set possible? What if the mean is five hours, the median is four hours, and the range is seven hours? Is more than one set possible? (Multiple sets are possible in each situation.) This activity helps the students understand that different data sets, with different distributions, can nevertheless share some summary information.

These activities prepare the students for the tasks on the new activity sheet, "Change the Data—Change the Stats?" Before distributing the sheet for the new exploration, say to your students, "You have examined the television viewing of a group of middle school students." Direct their attention again to the graph that they made of that data (see fig. 4.8), and pose questions such as the following:

- "What observations or conclusions can you make about these students' TV watching behavior?
- "How might you estimate the location of the median or the mean just by looking at the overall shape of the distribution of the data?"
- "What arguments could you give to justify your reasoning?"

See if your students remark on the fact that the majority of the data are located in the interval from 0 to 15 hours. Give each student a copy of the new sheet, "Change the Data—Change the Stats?" Point out that the graph on this sheet (see fig. 4.9) shows the data in this interval only. It "filters out" the six data values in the interval from 20 to 30 hours, interpreting those values as *outliers*. The graph has been electronically produced with the computer applet Plop It!

Ask your students to estimate the locations of the mean and the median for the revised distribution. They can also locate the mode, although this activity does not address this third measure of center. Be sure to have your students give reasons for their decisions to locate the other two measures as they did.

Explore

Assign your students to groups of two to four to complete the exploration on the activity sheet "Change the Data—Change the Stats?" If possible, have them work with the applet Plop It! If resources for using the technology are not available, give each group a sheet of chart paper and a pad of small sticky notes, enabling the students to make the graph on paper, using removable sticky notes to represent changing data.

"The use of computer tools [can be viewed] as an integral aspect of statistical reasoning rather than as technological add-ons."
(McClain, Cobb, and Gravemeijer 2000, p. 174)

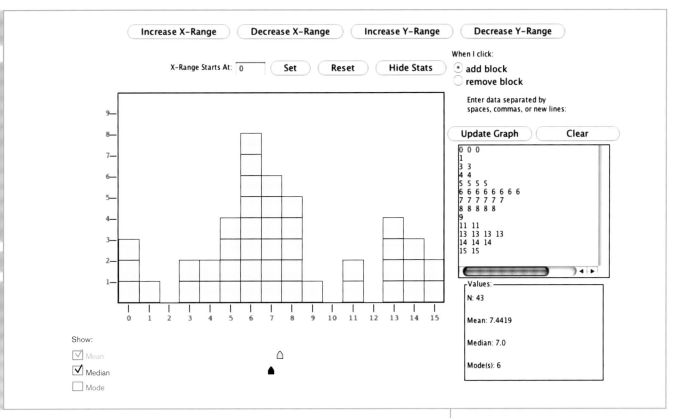

Fig. **4.9.**

A bar graph made with the applet Plop It! to show data from the activity TV Watching for the interval from 0 to 15 hours

The tasks on the new activity sheet reinforce learning from the earlier activities, TV Watching and Making the Data. The students begin their work by entering values from the television-viewing data set into the computer applet Plop It! (or on chart paper), reproducing the graph that they see on the new sheet. In steps 1–3, they then predict what will happen to the mean or the median if they change selected values in the data set. They enter the changes and check their predictions against the actual results, reasoning to explain what they thought would happen and what actually did happen.

Working in groups provides the students with opportunities for discussion and mutual support as they move through the tasks. If your students are working with the applet, you should also offer your own help as they learn how to enter data, remove data values ("remove a block"), "show stats," and so on. Because the activity focuses only on the mean and the median, the students can check the boxes next to the names of these two measures of center while leaving the box next to "mode" unchecked.

Students who are working with the applet will see that the software computes the measures of center, marks their locations and has the capability to reveal their values; students working with paper and sticky notes must do this work by hand. All students follow the instructions on the activity sheet to explore the impact of changes in the distribution of the values in a data set on its mean and median.

Step 4 calls on the students to clear the original data from the applet (or paper graph) and create a new distribution that meets certain requirements: the data set must include values on the television viewing of twelve people, both its mean and median must be 5 hours, and its maximum value—the greatest amount of time spent watching TV by any of the twelve people—must be 8 hours. The students then alter their new distribution in various ways and examine the effects on the mean and median.

Step 5—the last step in the activity—shows the students a graph with a new distribution. The students must estimate its median and mean without making any computations and explain their reasoning. When all the students have finished the tasks on the activity sheet, bring them together as a class to discuss their discoveries and their explanations.

Evaluate

Students usually have little difficulty in reasoning that the mean responds to changes in data. However, understanding when or why the median changes is often much more challenging. Thus, it is important to discuss the students' work on each of the questions carefully. In particular, the students can learn a great deal from the experience of looking at the different distributions that they created in step 4 and the estimates and predictions that they made in step 5. Such experiences can draw attention to the factors that influence the mean and the median of a distribution of data. Being able to make sense of a distribution and offer reasonable estimates of the locations of the mean and the median shows that students are growing in their abilities to reason statistically.

Extend

To reinforce your students' thinking about the mean and the median of a distribution and extend their independence in reasoning about these measures of center, encourage the students to continue to work with the data from the activity. Have them make different distributions that place the mean and median together, in the same location, or apart, in different ones. Ask them to exchange their distributions with other students, calling on them to estimate where these measures are located and explain their reasoning.

Conclusion

This chapter has highlighted the need for middle school students to have multiple opportunities to sort through and think about data to make decisions. Many and varied experiences in concrete contexts provide students with practice in determining what information to gather in particular contexts and what processes to use to analyze the information to answer particular questions.

Middle school students need to learn to be aware of their assumptions. They must understand that different perspectives are possible and that different processes of analysis can be equally valid and may lead to the same result. Above all, they must not lose sight of the concrete context in which a question for investigation arises, and they must recognize the importance of articulating their reasoning throughout the process of answering the question. Students should discover that those who ignore the context of the data may come to the same conclusion as those whose reasoning about the data is sound, but arguments that do not take the context into account have far less persuasive power.

"To support students' development of ways to reason logically about data, instructional tasks would need to build from the students' current understandings."
(McClain 1999, p. 379)

Navigating through Problem Solving and Reasoning in Grades 6–8

Data—regardless of the type—are the basis for many decisions. All the data that students reason about need not be numerical. Categorical data are often useful in making decisions as well. For example, in purchasing a car, the make, model, type of transmission, and even color can be just as important as price or fuel efficiency as measured in miles per gallon. Students can examine consumer guides to see what characteristics these sources use to rate various products and determine the top televisions, DVD players, microwave ovens, computers, and other goods.

The next chapter examines reasoning in another strand of mathematics—algebra. Many students have their first formal encounter with algebra in middle school. Because of the indispensable foundation that algebra provides for much of higher mathematics, algebraic reasoning is a topic of great importance.

PROBLEM SOLVING *and* REASONING

Chapter 5

Reasoning about Algebra

Preparing students for increasingly complex mathematics requires varied school experiences that foster essential habits of mind. Through meaningful and diverse mathematical experiences, students become accustomed to moving conceptually from specific to general, concrete to abstract, examples to rules. They learn to inspect mathematical objects purposefully, searching for common, recurring, or regularly changing characteristics, and they use their discoveries to make sense of the objects.

Such habits of mind give thinkers access to the underlying structures of mathematics. Algebraic reasoning most emphatically depends on their development. Blanton and Kaput (2005) define algebraic reasoning as "a process in which students generalize mathematical ideas from a set of particular instances, establish those generalizations through the discourse of argumentation, and express them in increasingly formal and age-appropriate ways" (p. 413). The authors continue, describing facets of this reasoning:

> Algebraic reasoning can take various forms, including (*a*) the use of arithmetic as a domain for expressing and formalizing generalizations (generalized arithmetic); (*b*) generalizing numerical patterns to describe functional relationships (functional thinking); (*c*) modeling as a domain for expressing and formalizing generalizations; and (*d*) generalizing about mathematical systems abstracted from computations and relations. (Blanton and Kaput 2005, p. 413)

"For students to generalize, they need to develop problem-solving schemas.... A schema allows similar experiences to be organized in such a way that the individual can easily recognize related experiences, assimilate and consolidate them and then quickly and easily retrieve the stored experiences when solving a related problem." (Steele 2005, p. 40; available on the CD-ROM)

This chapter focuses on the second category of algebraic reasoning identified by Blanton and Kaput—reasoning involved in "generalizing numerical patterns to describe functional relationships." This reasoning is essential in the creation of mathematical descriptions of the growth and change in patterns of objects, including numbers themselves (see, for example, the investigation Reasoning about Sums of Consecutive Numbers in chapter 1).

Experiences with patterns that continue and grow or change play vital roles in intermediate and middle-grades mathematics curricula. These experiences help establish habits of mind that allow algebraic reasoning to develop step by step, progressing to complex reasoning about functions.

The real power of work with visual patterns in middle school derives from the fact that this work allows students to move smoothly, without undue frustration, from concrete examples to general rules that use numbers or other symbols to describe perceived change. Thus, the fundamental work of the mathematician becomes increasingly accessible to the students. The process of discovering shared, recurring, or regularly changing features is natural, and the process of developing generalizations or function rules grows seamlessly out of the discoveries, giving students an important and reassuring model for algebraic reasoning in general.

Reasoning about Growth and Change in Patterns

Goals

- Analyze a growing pattern, moving from concrete examples to general rules about the pattern at any stage
- Use a table as a tool for organizing data and facilitating the development of algebraic expressions to describe the growth of the pattern
- Use generalizations obtained by examining the structure of a pattern to solve problems

Materials and Equipment

For each student—
- A copy of each of the following activity sheets:
  - "Growing in Front of Your Eyes"
  - "Change the Pattern—Change the Growth?"
  - "What Changes? What Stays the Same?"
- (Optional) One or two sheets of quarter-inch grid paper or a collection of small square tiles in two colors for each group of two or three students

For selected groups of two or three students—
- A sheet of chart paper or several blank overhead transparencies
- Colored markers

Prior Knowledge

Students' earlier experiences should have provided multiple opportunities to work with growing patterns (see, for example, the activities Snakes and More Snakes in *Navigating through Algebra in Prekindergarten–Grade 2* [Greenes et al. 2001, pp. 13–15, 75] and Watch Them Grow in *Navigating through Algebra in Grades 3–5* [Cuevas and Yeatts 2001, pp. 12–14, 74]). Activities in chapter 1 of *Navigating through Algebra in Grades 6–8* (Friel, Rachlin, and Doyle 2001) can serve as a preassessment or review of students' learning from beginning experiences with patterns.

Learning Environment

The students work in groups of two or three with the teacher serving as a facilitator and questioner, supporting them as they reason mathematically to generalize about the patterns in the activity.

Discussion

Experiences with growing patterns can be very effective in building middle school students' skills in algebraic reasoning. Such patterns often

pp. 168–72, 173–77, 178–81

The template "Quarter-Inch Grid Paper" on the CD-ROM allows you to print out sheets of grid paper for your students' use in the investigation.

"By using tables, charts, physical objects, and symbols, students [in the higher elementary grades] make and explain generalizations about patterns and use relationships in patterns to make predictions." (Friel, Rachlin, and Doyle 2001, p. 2)

Having students make growing patterns based on the different shapes of their first initials is an activity that Navigating through Algebra in Grades 6–8 *(Friel, Rachlin, and Doyle 2001, p. 10) suggests.*

have a geometric context; students can examine the patterns' physical, visible structures and organize numerical information about them in tables. Growing patterns that can be developed through the use of geometric models offer opportunities for students to reason about variables and relationships among two or more variables. *Navigating through Algebra in Grades 6–8* (Friel, Rachlin, and Doyle 2001, pp. 7–8) identifies many sorts of reasoning tasks in which experiences with sequences of shapes typically engage students:

- finding the first few terms of the sequence and recording this information in a table or chart;

- drawing the next shape in the sequence;

- describing the shapes succinctly with words in such a way that someone who has not seen them will be able to duplicate the sequence;

- writing the rule that will produce the ever-growing sequence with the first few terms as given (either a recursive or an explicit form of pattern generalization);

- generalizing to make predictions for the tenth, the forty-third, or the *n*th shape in the sequence;

- comparing different ways of arriving at the generalization (e.g., equivalence of algebraic expressions or recursive and explicit forms of pattern generalization).

Engage

Ask your students to suppose that a middle school student named Laura has created a growing pattern based on her first initial, as shown in figure 5.1. In designing her pattern of Ls, Laura has had to make some decisions. See if your students can explain how she has made her pattern grow.

Fig. **5.1.**

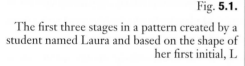

The first three stages in a pattern created by a student named Laura and based on the shape of her first initial, L

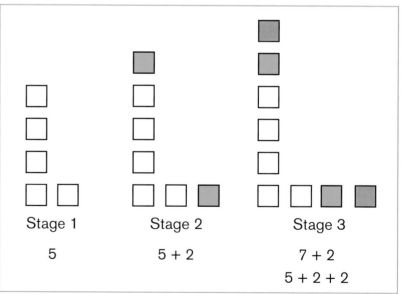

At stage 1, Laura used two tiles in the base of the L and four on the side, with the "corner" tile serving as part of both the base and the side. Because this tile doubles as a base and a side tile, stage 1 of Laura's

pattern has a total of five tiles instead of six. To form stage 2, Laura added two tiles—one to the open end of the base, and one to the open end of the side. The resulting letter L has three tiles on the base and five tiles on the side. Adjusting for the double-counted vertex tile gives a total of seven tiles at stage 2, which, like stage 1, has two more tiles on the side than the base. Thinking about the total number of tiles at stage 2 in terms of the total tiles at stage 1 plus the two new tiles leads to the numerical expression 5 + 2, or 7, tiles for the total in stage 2. Laura added two tiles to the base and side of stage 2 to create stage 3. Thus, stage 3 has the five tiles at stage 1, plus the two additional tiles to make stage 2, plus two new tiles, or nine tiles in all: 5 + 2 + 2 = 9.

Helping students examine the physical structure of a growing pattern in this way prepares them for making predictions about a later stage. Can your students tell how many tiles Laura will need for stage 10, for example? See if they can reason that the L at this stage consists of 5 + (2 × 9), or 23, tiles. The base has 10 + 1, or 11, tiles, and the side has two more tiles, or 13 tiles. Adjusting for the "double counting" of the vertex tile gives a total of 23 tiles at stage 10.

Analyzing the physical structure of a pattern can help students make sense of a way to predict a pattern's growth and generalize about its structure at stage n. In the case of Laura's pattern, the students can see that stage n consists of $5 + 2(n - 1)$ tiles. Moreover, they can understand that this expression lets them find the number of tiles whenever they know n, the stage of an L in the sequence.

Recording and organizing information about a growing pattern in a table can be a very useful way for students to make sense of its change or growth. Laura might use a three-column table such as that in figure 5.2 to help her understand the growth in her pattern. The numerical descriptions in the second column provide a picture of the growth.

Stage	Numerical Description of the Pattern	Total Tiles
1	5	5
2	5 + 2	7
3	5 + 2 + 2	9
4	5 + 2 + 2 + 2	11
5	5 + 2 + 2 + 2 + 2	13
⋮	⋮	⋮
n	$5 + 2(n - 1)$	$2n + 3$

Fig. **5.2.**

A three-column table showing the numbers of tiles at different stages of Laura's growing pattern of Ls

Data in a table are removed—*abstracted*—from the pattern under examination. Working with the values in the table often assists students in making the conceptual step from specific to general. It can help them notice important mathematical relationships. It is worth remembering, however, that working with the values in this way becomes a disadvantage if it causes students to lose sight of the concrete context and unique characteristics of the pattern itself. Constantly making connections between the information recorded in a table and a pattern's physical structure supports students' developing understanding of the process of generalizing and making predictions about a pattern.

Explore

Give each student a copy of the activity sheet "Growing in Front of Your Eyes." The sheet presents the first three stages in a growing pattern (used in research by Cai and Hwang [2002]). At each stage, a "border" of white square tiles surrounds a square "core" of black square tiles. Figure 5.3 shows stages 1–3 in the pattern.

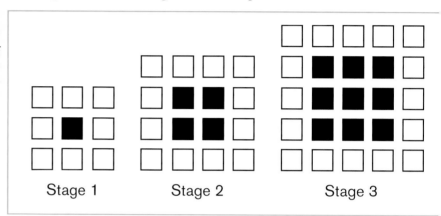

Fig. **5.3.**

The activity sheet "Growing in Front of Your Eyes" shows the first three stages in a growing pattern of black and white square tiles.

Assign the students to groups of two or three so that they can work collaboratively on the questions on the activity sheet. These questions move the students from a concrete, visual exploration of the sequence to an abstract, algebraic analysis of its growth and then ask them to use what they have learned to solve problems related to the pattern. If you wish, also give each student one or two sheets of grid paper, or a collection of small square tiles in two colors, to aid in their investigations.

Step 1 starts the process by having the students draw stage 4. They explain why they need to draw particular numbers of black and white tiles at this stage.

In step 2, the students take a step toward conceptualizing the growth from stage to stage as they think about the numbers of black tiles, white tiles, and total tiles at stage 6. They stretch mentally from their physical drawing of stage 4, taking into account the growth that would occur at stage 5 and then at stage 6.

Step 3 introduces a table to help the students organize the information that they have been gathering about the sequence. They use combinations of words and numbers to describe the numbers of tiles (black tiles, white tiles, and total tiles) at stages 1–6. The growth in the numbers of black tiles and total tiles is likely to be easier for the students to describe than the growth in the numbers of white tiles. Figure 5.4 shows a sample completed table that illustrates one way of describing this growth; students may offer other descriptions, some of which are discussed later. No matter how the students describe the growth, inspecting the information that they enter in the table can assist them in understanding the basic characteristics of the pattern and the relationships among the characteristics.

Stage	Number of Black Tiles–Describe What You See	Number of White Tiles–Describe What You See	Total Number of Tiles–Describe What You See
1	A 1 × 1 square made up of 1 black tile	A square border made up of (3 × 2) + (1 × 2) = 6 + 2 = 8 white tiles	A 3 × 3 square made up of 9 tiles in all
2	A 2 × 2 square made up of 4 black tiles	A square border made up of (4 × 2) + (2 × 2) = 8 + 4 = 12 white tiles	A 4 × 4 square made up of 16 tiles in all
3	A 3 × 3 square made up of 9 black tiles	A square border made up of (5 × 2) + (3 × 2) = 10 + 6 = 16 white tiles	A 5 × 5 square made up of 25 tiles in all
4	A 4 × 4 square made up of 16 black tiles	A square border made up of (6 × 2) + (4 × 2) = 12 + 8 = 20 white tiles	A 6 × 6 square made up of 36 tiles in all
5	A 5 × 5 square made up of 25 black tiles	A square border made up of (7 × 2) + (5 × 2) = 14 + 10 = 24 white tiles	A 7 × 7 square made up of 49 tiles in all
6	A 6 × 6 square made up of 36 black tiles	A square border made up of (8 × 2) + (6 × 2) = 16 + 12 = 28 white tiles	An 8 × 8 square made up of 64 tiles in all
⋮	⋮	⋮	⋮
n	An $n \times n$ square made up of n^2 black tiles	A square border made up of $((n + 2) \times 2) + (n \times 2)$ = $2n + 4 + 2n$ = $4n + 4$ white tiles	An $(n + 2) \times (n + 2)$ square made up of $n^2 + 4n + 4$ tiles in all

Fig. 5.4.

A sample of a completed table organizing information on black tiles, white tiles, and total tiles in stages 1–6 and n

Equally important to the students' development of algebraic thinking is gaining an understanding of the usefulness of mathematical descriptions of these characteristics and relationships in the general case—the case of stage n. The last row in the table calls for expressions in terms of n. Be sure that your students move back and forth between a detailed visual inspection of the pattern and an examination of the mathematical values in the table, confirming any observations drawn from the table by observations drawn from the stages themselves, and vice versa. Students who think concretely and make detailed observations should not find the task of coming up with the algebraic expressions to be excessively difficult. You might also ask your students to use grid paper and make graphs to assist them in visualizing the relationships in another way.

Steps 4–6 invite the students to apply what they have learned about the sequence. In step 4, the students make important observations about the "big picture" by reasoning from what they know about black tiles at different stages. They conclude that odd-numbered stages always have odd numbers of black tiles, and even-numbered stages always have even numbers of black tiles. They use this information to predict that stage 17 has an odd number of black tiles.

In step 5, they use their discoveries to determine what stage in the sequence has 44 white tiles. How many black tiles, and how many tiles in all, does this stage have? In step 6, they consider whether any stage in the sequence has 52 black tiles. Does any stage have 52 white tiles or 52 tiles in all?

Give your students sufficient time to think carefully about the questions and work without rushing. Their conversations can give you insight into their reasoning, so move around the classroom and listen, supplying guidance and direction as needed. Select groups of students to write their approaches to particular problems on chart paper or transparencies to share afterward with the class. When all the students have completed the questions or made as much progress as possible, bring them together for a class discussion of their work.

Evaluate

The students should have no trouble with the concrete preliminaries that move them gently toward abstract and general thinking about the three basic elements of the stages in the pattern—black tiles, white tiles, and total tiles (black and white tiles together). After drawing stage 4 in step 1 and imagining stage 6 in step 2, they should find themselves on their way to an algebraic understanding of the components of stage n, with n as any natural number. They are likely to reason successfully in step 2 that stage 6 has 36 black tiles, 28 white tiles, and 64 tiles in all. Nevertheless, explaining why they think so may present some challenge especially in the case of the white tiles.

Figure 5.5 shows the reasoning of four students in support of their conclusion that stage 6 has 28 white tiles. The student providing explanation (*a*) reasoned in the way that the table in figure 5.4 illustrates: stage 6 would have 6 + 2, or 8, tiles on each of two opposite sides, and 6 tiles on each of the other two opposite sides, for a total of 28 tiles. The students providing explanations (*b*)–(*d*) all reasoned in another way to reach the same conclusion. They found the number of white tiles in stage 6 as the difference between its total number of tiles and its number of black tiles. Furthermore, the student offering explanation (*d*) extended this reasoning to the general case of stage n.

Generalizing about stage n, with n as any natural number, is an important prerequisite for efficient problem solving related to the pattern. In terms of n, how many black tiles, white tiles, and total tiles are in the pattern at stage n? Students who enter detailed information about stages 1–6 in the table in step 3 are in a good position to perform a successful analysis of the basic components of the pattern and the relationships that determine its growth from one stage to the next.

Allow your students to share all their discoveries about how the pattern of tiles grows from stage to stage. Pay close attention to the ways in which they are thinking about the principal elements of the pattern and the ways in which these characteristics change and grow. Do the students offer correct and meaningful mathematical descriptions of the changes? Do they understand the descriptions and their relationships to the components? Often students do not think with sufficient

(a) 28, because there are 8 eight squares along and the bottom and 6 uncounted ones on each side. $(8 \times 2) + (6 \times 2) = 28$

(b) There would be 28 white squares because it would be an 8 by 8 square. When multiplied, That's 64 squares. 36 of the squares are black. 36 from 64 is 28.

(c) There are 28 white squares in the sixth figure. I know this because the number of outside tiles is equal to the number of total tiles minus the number of center tiles. $(6+2)^2 = 64$. $64 - 36 = 28$.

(d) There are 28 white tiles in the 6th figure. To figure this out you would take the figure number add two, then square that number. then subtract the figure number squared.

$$[(6+2)^2] - 6^2 = 28$$

Fig. **5.5.**

Responses from four students to two questions: "How many white tiles does stage 6 have?" and "Why do you think so?"

detail or specificity about the elements of a pattern, so focusing on these three elements in your discussion can be helpful.

1. *Black tiles.* At each stage, the black tiles form a square inside a one-tile-wide "frame" of white tiles. The growing numbers of black tiles composing these squares form the sequence of square numbers: 1, 4, 9,... (the squares of the natural numbers 1, 2, 3, ...). Figure 5.6 shows the first three stages in the pattern of the black "cores." Your students have probably recognized without much difficulty that the number of black tiles at stage n, where n is any natural number, is n^2.

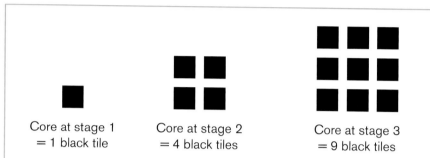

Core at stage 1
= 1 black tile

Core at stage 2
= 4 black tiles

Core at stage 3
= 9 black tiles

2. *White tiles.* Students can consider the number of white tiles in many different ways. Let your students present all the methods that they used. Descriptions of several approaches follow.

   a. Students can think about the white tiles in two groups—corner tiles and tiles between corners (see fig. 5.7). They can observe that every stage has 4 corner tiles. On each side, the number of white tiles between the corners is equal to n, the number of the stage. (Each side of stage 1 has 1 tile between corners, each side of stage 2 has 2 such tiles, each side of stage 3 has 3, and so on.) They can conclude that on all four sides, stage n has $4 \times n$ white tiles between corners. By adding to this number of tiles the 4 tiles at the corners, the students obtain a total of $4n + 4$ white tiles at stage n.

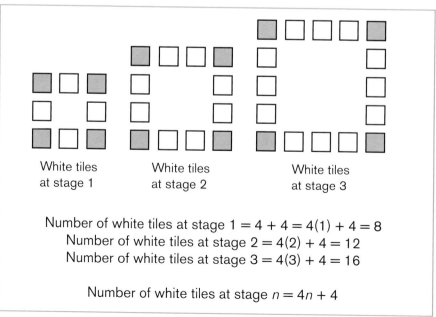

White tiles
at stage 1

White tiles
at stage 2

White tiles
at stage 3

Number of white tiles at stage 1 = 4 + 4 = 4(1) + 4 = 8
Number of white tiles at stage 2 = 4(2) + 4 = 12
Number of white tiles at stage 3 = 4(3) + 4 = 16

Number of white tiles at stage $n = 4n + 4$

b. Students can separate the white tiles at any stage into four equal groups, with one corner tile in each group (see fig. 5.8). They can conclude that at stage *n* each group has *n* + 1 tiles, so in all, stage *n* has 4(*n* + 1), or 4*n* + 4, white tiles.

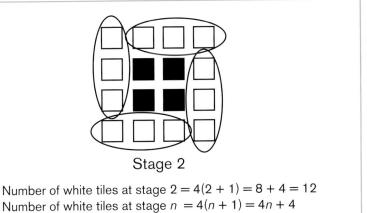

Stage 2

Number of white tiles at stage 2 = 4(2 + 1) = 8 + 4 = 12
Number of white tiles at stage *n* = 4(*n* + 1) = 4*n* + 4

Fig. **5.8.**

Separating the white tiles at any stage into four equal groups, each with one corner tile

c. Students can apply what they know about finding the perimeter of a square to evaluate the white tiles in the one-tile-wide "frame" around the black "core" at each stage. A square's perimeter is equal to its side length times 4. The students can approach the one-tile-wide "frame" of white tiles as the "perimeter" of a square with a side length of *n* + 2 tiles *if* they recognize that they must adjust for the width of the frame (see fig. 5.9). They should be able to see that a count of 4 × (*n* + 2) white tiles "doubles up" the corner tiles in the same way that Laura's tally of the tiles in an L double-counted the tile at the vertex of the L's stem and base. When the students subtract the double-counted corners, they discover that that at any stage the white-tile frame consists of 4(*n* + 2) – 4, or 4*n* + 8 – 4, or 4*n* + 4, tiles.

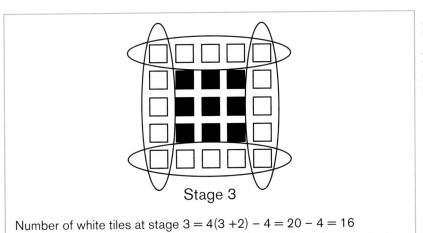

Stage 3

Number of white tiles at stage 3 = 4(3 + 2) – 4 = 20 – 4 = 16
Number of white tiles at stage *n* = 4(*n* + 2) – 4 = 4*n* + 8 – 4 = 4*n* + 4

Fig. **5.9.**

Approaching the one-tile-wide "frame" of white tiles at any stage as the perimeter of a square that is *n* + 2 tiles on a side involves adjusting for the double-counted corners.

d. The students can use the method shown in the table in fig. 5.4. They can divide the white tiles in the "frame" of any stage into two groups—one consisting of opposite sides "with corners" and the other consisting of opposite sides "without corners" (see fig. 5.10). The students can reason that each side with corners consists of (*n* + 2) tiles, and each side without corners consists of *n* tiles, so the count of white tiles at stage *n* is $2(n + 2) + 2n$, or $2n + 4 + 2n$, or $4n + 4$, tiles.

Fig. **5.10.**

Sorting the white tiles at any stage into two sets, one composed of opposite sides "with corners" and the other composed of opposite sides "without corners"

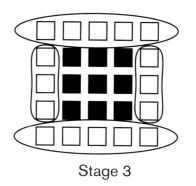

Stage 3

Number of white tiles at stage 3 = $2(5) + 2(3) = 10 + 6 = 16$
Number of white tiles at stage *n* = $2(n + 2) + 2n = 2n + 4 + 2n = 4n + 4$

> *e.* Many students—perhaps the majority—are likely to think of the number of white tiles as the difference between the total number of tiles and the number of black tiles. This approach is discussed in greater detail below.

3. *Total tiles.* Students can find the total number of tiles at any stage as the sum of its black and white tiles. Adding the number of black tiles, n^2, and the number of white tiles, $4n + 4$, gives a total of

$$n^2 + 4n + 4, \text{ or } (n + 2)^2,$$

tiles at stage *n*. However, students can also find—and perhaps are more likely to find—the number of total tiles at any stage irrespective of the numbers of black and white tiles. Ignoring the colors of the tiles, they may simply consider each stage as a square composed of undifferentiated tiles. In this process, they observe that the number of tiles on a side of the square is always 2 more than the number of the stage. That is, stage 1 has $1 + 2$, or 3, tiles on a side; stage 2 has $2 + 2$, or 4, tiles on a side; stage 3 has $3 + 2$, or 5, tiles on a side; and so on. Thus, each stage has $(n + 2)^2$, or $n^2 + 4n + 4$, tiles in all, counting black and white tiles together. Indeed, students may notice the simple squaring in total tiles and black tiles before they think about the number of white tiles as a separate quantity. These discoveries will allow them to find the number of white tiles at stage *n* easily, as the difference between the total number of tiles and the number of black tiles:

$$(n + 2)^2 - n^2 = n^2 + 4n + 4 - n^2 = 4n + 4.$$

As you continue to discuss your students' work, assess how skillful they are in applying their reasoning and conclusions in step 3 to the problem solving that steps 4–6 call on them to do. Evaluate their solutions and explanations.

In step 4, the students should conclude that stage 17 has an odd number of black tiles. Can they justify this conclusion by what they have discovered about the numbers of black tiles at stage *n*? See if they are able to build an argument on their discovery that the number of black tiles is equal to the square of *n*. The number of black tiles at stage 17 is odd because of three things: n is odd at stage 17, the number of black tiles at stage n is n^2, and the product of odd numbers is odd: $17^2 = 289$.

Step 5(a) asks the students to determine at what stage the pattern has 44 white tiles. Students who have successfully completed the table in step 3 and have had some experience in solving equations can make direct and efficient use of their observations about the growth of white tiles in the pattern and their discovery that stage n has $4n + 4$ white tiles. They can set up an equation and solve for n:

$$4n + 4 = 44$$
$$4n = 44 - 4$$
$$4n = 40$$
$$n = 10.$$

Stage 10 has 44 white tiles.

Even without experience in setting up and solving equations in one variable, students who have determined that the number of white tiles at stage n is equal to $4n + 4$ can use this expression to solve the problem fairly quickly by guessing and checking: $4n + 4$ equals 44 when n equals 10, so the stage that has 44 white tiles is stage 10.

Students who are less confident in using symbolic expressions can still reason about the problem and solve it by applying their observations about the growth in white tiles from stage to stage. Those who have approached the growth by separating the white tiles at each stage into corner tiles and tiles between corners (see fig. 5.7) can reason as the student did in the following solution: "If there are 44 white tiles, this means that 4 tiles are corner tiles, so $44 - 4 = 40$ white tiles that are the side tiles. There are 4 sides with the corners removed, so $40 \div 4 = 10$. The number of tiles on a side, not counting corner tiles, is equal to the number of the stage. Stage 10 has 44 white tiles."

Figure 5.11 shows another student's solution and explanation of his work on the problem. This student found the number of the stage

> It is the 10th figure. To figure this out you would divid! the outside (white) tiles by four (for the four corrners) then minus one. $\frac{44}{4} - 1 = 10$

Fig. **5.11.**

A student's explanation of his work in solving the problem in step 5: "Suppose that a particular stage in the pattern has 44 white tiles. What is n, the number of the stage?"

by dividing the number of white tiles by 4 and then subtracting 1:

$$\frac{44}{4} - = 10$$

Thus, it appears that the student reasoned about the growth in the white tiles by separating the tiles into four equal groups, as in figure 5.8, and then relating the size of a group to the stage. The number of white tiles in each group is one more than the number of the stage; subtracting 1 thus gives the stage. (In the case of stage 1, which has 8 white tiles,

$(8 \div 4) - 1$ equals 1; in stage 2, which has 12 tiles, $(12 \div 4) - 1$ equals 2; and so on.)

The student's calculation and result imply that he reasoned in this way, but his explanation suggests that his grasp of the ideas was not as firm as it have should been. It is possible that his vague and seemingly incorrect reason for dividing by 4—"for the corners"—reflects a lack of skill in describing his process adequately rather than a failure of understanding. Further probing might have revealed the extent to which the student actually understood the relationship between the number of white tiles and the stage in the pattern. In any case, discussion might have strengthened his thinking and his ability to articulate the details of his solution process.

Another student solved the problem by making a chart in which she paired stages with their respective numbers of white tiles until she reached 44 white tiles (see fig. 5.12). As the student explained, this solution process depends on knowing that the numbers of white tiles grow by 4 from one stage to the next. This method has its limitations—its success depends on having a number of white tiles that is small enough to make the method feasible. Although the student apparently was able to express the number of white tiles at stage n algebraically, writing "$4n + 4$" to the right of her chart, she may not have known how to use this expression to solve the problem more efficiently.

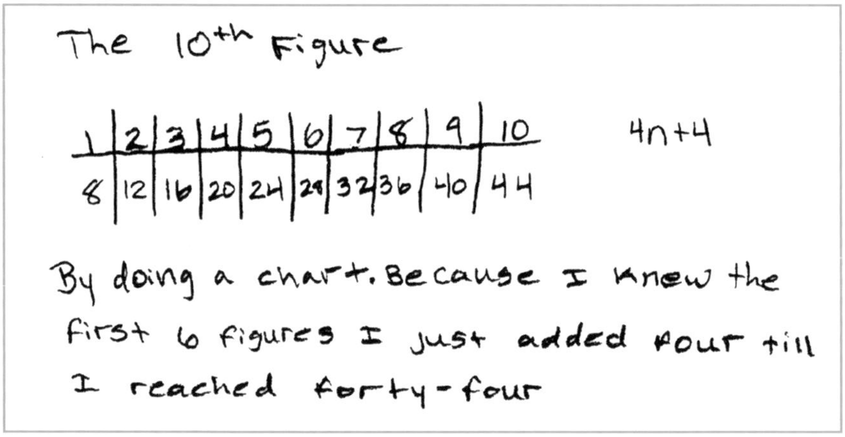

Fig. **5.12.**

A student's use of a chart to solve the problem in step 5: "Suppose that a particular stage in the pattern has 44 white tiles. What is n, the number of the stage?"

No matter how students reason, all who have correctly determined that the stage with 44 tiles is stage 10 should be able to solve parts (*b*) and (*c*) of step 5 easily. See if your students understand that they can simply square the number of the stage to find the number of black tiles: $10^2 = 100$ black tiles. They can then use simple addition in part (*c*): $44 + 100 = 144$ tiles in all at stage 10.

Part (*a*) of step 6 asks the students if any stage can have a total of 52 tiles (black and white tiles together). The students' inspection of their data in the table in step 3 should have shown them that the total number of squares is always a square number. The number 52 is not a square. Moreover, the students should note that in the table the total number of tiles goes from 49 tiles at stage 5 to 64 tiles at stage 6. See if your students reasoned the same way in part (*b*) to argue that no stage has 52 black tiles.

The students may have found the question in part (*c*) more challenging: "Is it possible for any stage to have 52 white tiles?" As in part (*a*) of step 5, if students reasoned successfully that stage n has $4n + 4$ white tiles, and they know how to set up and solve an equation in one variable, they may be able to solve this problem efficiently:

$$4n + 4 = 52$$
$$4n = 48$$
$$n = 12.$$

Students who are less advanced in algebraic thinking and less familiar with algebraic processes can use one of the other methods discussed for part (*a*) of step 5. If they know that $4n + 4$ is an expression for the number of white tiles in stage *n*, they can use a guess-and-check approach to see whether they can find a value for *n* that gives them 52 white tiles. Or they can extend the table to see if 52 is a value for the number of white tiles in the row for some stage in the pattern.

Extend

Students can extend the exploration in the activity sheet "Growing in Front of Your Eyes" by working with one or both of the activity sheets "Change the Pattern—Change the Growth?" and "What Changes? What Stays the Same?" The first sheet builds directly on "Growing in Front of Your Eyes." The second sheet follows more closely on the "Engage" section of the investigation, which presents Laura's growing pattern of Ls.

In the case of either extension, give each student a copy of the activity sheet and have the students work collaboratively as before, in small groups. If you wish, supply each group with one or two sheets of grid paper. Select groups to present their work on chart paper or overhead transparencies for the whole class to see and discuss when everyone has finished working on the problems on the activity sheet.

"Change the Pattern—Change the Growth?" The first activity sheet presents the students with a modified version of the growing pattern in "Growing in Front of Your Eyes." Figure 5.13 shows stages 1–3 in the new pattern. Note that this pattern removes two corner tiles—white tiles—from opposite corners at each stage, along with the two white tiles that neighbor each of these corner tiles. Hence, each stage has 6 fewer white tiles than its counterpart in the original sequence. Thus, the number of white tiles in stage *n* goes from $4n + 4$ in the first pattern to $(4n + 4) - 6$, or $4n - 2$, in the second. At the same time, the number of black tiles at stage *n* remains the same as before, or n^2, while the number of total tiles goes from $n^2 + 4n + 4$, or $(n + 2)^2$, to $n^2 + 4n - 2$.

The format of the new activity sheet parallels that of "Growing in Front of Your Eyes." Again, the students move from very concrete

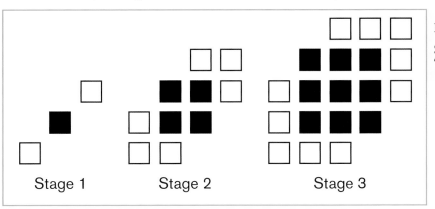

Stage 1   Stage 2   Stage 3

Fig. **5.13.**

Stages 1–3 in the modified pattern in "Change the Pattern—Change the Growth?"

work with the pattern to tasks that call for increasingly abstract and complex algebraic thinking. As before, a table is at the center of the activity, facilitating algebraic reasoning about relationships among the stage in the pattern and the numbers of black tiles, white tiles, and black and white tiles together. After examining these relationships, the students solve problems as on the earlier sheet.

"What Changes? What Stays the Same?" Most of the growing patterns that the second activity sheet presents are based on the shapes of letters of the alphabet. This extension thus recalls Laura's sequence of Ls and continues to focus the students' attention on physical models as sources of information about the structures of patterns. However, the new activity sheet introduces a variation that is useful in stretching students' reasoning. Instead of giving the students the first three stages in a pattern, this sheet omits stages 1 and 3, showing only stage 2, or it omits stages 1 and 2, showing only stage 3. Thus, the students must reason back to earlier stages in a pattern, as well as ahead to later ones. In this way, the problems offer opportunities for the students to exercise *reversibility* in reasoning.

You might approach step 1 as a whole-class exercise. The illustration on the activity sheet shows stage 2 in a growing pattern of Hs (see fig. 5.14). The students can see that the H at stage 2 has 10 tiles, and step 1(*a*) gives them the information that the H at stage 1 has 9 tiles and the H at stage 3 has 11 tiles. Thus, the Hs grow from stage to stage by adding just one tile. Because of the symmetry of the letter H, the students can reason that only the "cross bar" of the H changes and grows from stage to stage.

Fig. **5.14.**

Stage 2 in a pattern of Hs; if stage 1 has 9 tiles and stage 3 has 11 tiles, only the "cross bar" grows from stage to stage.

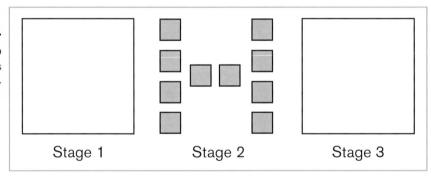

You can also ask the students what stages 1 and 3 would look like if stage 1 had 5 tiles and stage 3 had 15 tiles. (In this case, each new stage would add a tile to the cross bar *and* to the top and bottom of each side of the H, for a total of 5 additional tiles each time.) By posing such a question, you can make students aware of the physical structure of the emerging pattern by considering *where* an H adds tiles, not just *how many* tiles it adds in all, as the pattern grows from stage to stage.

To extend all this work with growing patterns yet further, you might have your students create patterns of their own and pose problems related to them that are similar to those on the activity sheets. For example, a student might present stage 3 in a pattern and give other students clues that allow them to draw stages 1 and 2, predict the number of tiles at a much later stage, or write a formula for the number of tiles at stage *n*. If the students come up with different numbers or configurations for earlier or later stages, they can discuss their reasoning about what stayed the same and what changed from stage to stage in the patterns.

Conclusion

Driscoll (1999) highlights the importance of helping students develop rules to represent functions: "Critical to algebraic thinking is the capacity to recognize patterns and reorganize data to represent situations in which input is related to output by well-defined functional rules" (p. 2). This chapter has emphasized the value of having students work closely with a growing pattern and connect its elements in very concrete and physical ways with any mathematical description or function rule that they develop for the pattern. As students become more comfortable with these ideas, they can move from visual, physical contexts to those that are exclusively abstract—for example, contexts in which information is presented only in a table and the students must develop rules from patterns that are strictly numerical.

In any case, the habit of looking beyond a perceived pattern to determine what "always works" and express this discovery as a mathematical rule about the pattern involves an ability to generalize that reflects a broader capacity of mathematical thinking. The next chapter explores a context in which middle school students can extend their reasoning abilities to a generalized proof.

NAVIGATIONS SERIES

GRADES 6-8

PROBLEM SOLVING *and* REASONING

Chapter 6

Reasoning and Proof, Levels 1–3

As the introduction to this book discussed, Waring (2000) identifies six levels in the development of students' understanding of proof. As students grow in mathematical sophistication, they progress gradually from level 0, characterized by a complete unawareness of any need for proof, toward level 5, evidenced by an understanding not only of the necessity and nature of proof but also of ways to construct proofs in diverse contexts. Almost every student moves beyond level 1, but not every student reaches level 5. Figure 0.5 (see p. 7) provides an overview of Waring's level-by-level articulation of students' understanding of proof.

Students in the middle grades do much of their reasoning at levels 1 and 2 of this framework. Students at level 1 have discovered that some conjectures need proof but think that finding a few supportive examples demonstrates their truth conclusively. Students at level 2 recognize that a few examples are not sufficient to prove a conjecture but are content to think that they simply need to find either additional examples that are more varied or random or a more general example as a proof for a class.

However, students' understanding of proof in grades 6–8 need not stop at level 2. Middle school mathematics affords numerous opportunities for students to advance to level 3. Students who attain this level of understanding are aware of the need for a generalized proof of a conjecture, although they are as yet unable to construct a valid proof by themselves. Nonetheless, they are likely to understand a proof at an appropriate level of difficulty, and they can often follow an explanation of the steps in the construction of such a proof. This chapter offers a path to level 3 that takes students through an investigation of the Pythagorean theorem.

Reasoning about the Pythagorean Theorem

Goals

- Explore the Pythagorean theorem to move from level 1 to level 3 in Waring's (2000) developmental framework of understanding mathematical proof

- Gain a well-reasoned understanding of the Pythagorean theorem by understanding a general proof of the theorem

Materials and Equipment

For each student—

- A copy of each of the following activity sheets:
  - "Take a Look at Those Squares"
  - "Sets of Right Triangles for 'Can You Prove It?'"
  - "Can You Prove It?"
- Three or four sheets of centimeter grid paper

For each group of two to four students—

- (Optional) A ruler calibrated in centimeters and millimeters
- A calculator
- A pair of scissors
- Access to the applet Squaring the Triangle (available on the CD-ROM and on the Web at http://www.shodor.org/interactivate/activities/pyth/index.html)

For the teacher—

- (Optional) A sheet of chart paper or one or two overhead transparencies

pp. 182–188, 189, 190–91

The template "Centimeter Grid Paper" on the CD-ROM allows you to print out sheets of grid paper for your students' use in the investigation.

Prior Knowledge

The students should know how to find the areas of a square and a triangle and have had previous experience in decomposing shapes into other familiar shapes (see chapter 2). Students will also find it helpful to have worked with area models for multiplication and the distributive property of multiplication over addition.

Learning Environment

The students complete their own activity sheets while working together in pairs, with the teacher encouraging them to formulate conjectures, evaluate them, and eventually attempt to justify those that they suppose to be true. The teacher serves as a facilitator and question-er throughout, supporting students as they reason about the relation-ship among the squares constructed on the sides of a right triangle.

If your classroom has only one computer but you can project the screen image for everyone to see, you can use the applet in a whole-class demonstration instead of having the students work with it in small groups. (The investigation can also be effective without the applet.)

Discussion

The Pythagorean theorem states that the sum of the areas of squares constructed on the legs of a right triangle is equal to the area of a square constructed on the triangle's hypotenuse. Figure 6.1 illustrates the theorem and shows an algebraic expression of the relationship between the area of the square on the hypotenuse and the areas of the squares on the legs of the triangle.

An exploration of this theorem can move students toward a general proof, taking them from level 1 to level 3 in their thinking, as described in Waring's (2000) framework:

> Level 1 — Students are aware of the need to prove a conjecture but think that finding a few examples that support it is enough for a proof.
>
> Level 2 — Students are aware that finding a few examples that support a conjecture does not prove it, but they think that examples that are more varied or selected more randomly provide proof or that a generic example forms a proof for a class.
>
> Level 3 — Students are aware of the need for a generalized proof of a conjecture, although they are as yet unable to construct such a proof on their own. Nevertheless, they are likely to understand simple proofs and follow explanations of their construction.

Engage

Ask, "What is a right triangle?" Review the fact that a right triangle has a right angle, which measures 90 degrees. Remind the students that the side that is opposite the right angle is called the *hypotenuse*. Elicit the fact that the triangle's other two angles are *acute*—their measurements are less than 90 degrees—and the sides that are opposite these angles are called the *legs* of the right triangle.

On the board, draw a right triangle with squares on each of its sides. Ask, "Do you suppose that any special relationship exists among the areas of these three squares?" Many students will suspect (or have already learned) that the area of the square on the hypotenuse is equal to the sum of the areas of the squares on the legs (see fig. 6.1). In this investigation, the students will explore this relationship.

Explore

Give each student a copy of the activity sheet "Take a Look at Those Squares." Assign the students in groups of two (up to four, if necessary) to work at classroom computers with the applet Squaring the Triangle, which comes into play in step 5. If you have only one computer in your classroom but can project a screen image for everyone to see, you can gather your students together for a whole-class consideration of this step. If you have no classroom computers at all, your students can still complete the activity sheet successfully by making right triangles on grid paper.

In steps 1–4, the students inspect different right triangles with squares constructed on their sides. They determine the areas of the squares (as well as their side lengths) and make conjectures about a relationship among the areas. To get your students started, you may

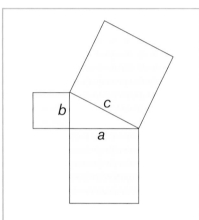

If—

- *c* is the length of the hypotenuse of a right triangle;

- *a* and *b* are the lengths of the legs of the triangle;

then $a^2 + b^2 = c^2$, with—

- c^2 as the area of a square with a side length of *c*;

- a^2 as the area of a square with side length *a*; and

- b^2 as the area of a square with side length *b*.

Fig. **6.1.**

According to the Pythagorean theorem, if the hypotenuse of a right triangle has a length of *c* and the triangle's legs have lengths of *a* and *b*, then $a^2 + b^2 = c^2$.

want to walk them through step 1. Depending on their level of mathematical understanding, they may need some guidance at the outset.

Note that all the right triangles in steps 1–4 appear on grids, with all vertices at the intersections of grid lines. In steps 1 and 2, the legs of the triangles lie on grid lines as well (see fig. 6.2). Thus, in these steps, the sides of the squares on the triangles' legs lie on grid lines, making the areas of these squares very easy to determine. The area of the square on the hypotenuse is another matter, however. Although this square's vertices are at the intersections of grid lines, its sides do not lie on grid lines, and thus finding the square's area is more problematic.

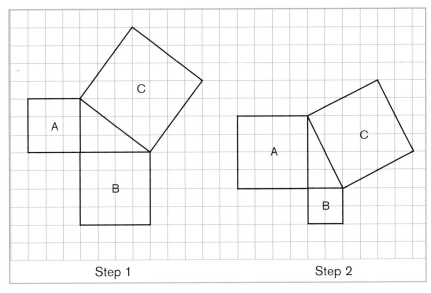

Step 1   Step 2

Fig. **6.2.**

In both step 1 and step 2, the sides of the squares on the legs of the right triangle lie on grid lines, but the sides of the square on the hypotenuse do not.

You can guide your students in decomposing this square along grid lines into four congruent right triangles and an enclosed square that is bounded on all four sides by grid lines (see fig. 6.3). Then they can easily find the area of the square on the hypotenuse by finding the areas of the component shapes that they have defined. To find the area of one of the triangles, they can apply the formula

$$A = \frac{1}{2} bh.$$

After the students have found the total area of the square on the hypotenuse by adding the areas of the four congruent right triangles

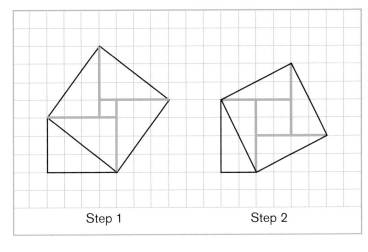

Step 1   Step 2

Fig. **6.3.**

The square on the hypotenuse of the right triangle in step 1 and step 2, decomposed along grid lines into four congruent triangles and an enclosed square, with all vertices at the intersections of grid lines and the enclosed square, they can find the side length of the larger square as the square root of this total area.

Fig. **6.4.**

Only one side of the right triangle in step 3 lies on a grid line, and no side of the right triangle in step 4 lies on a grid line.

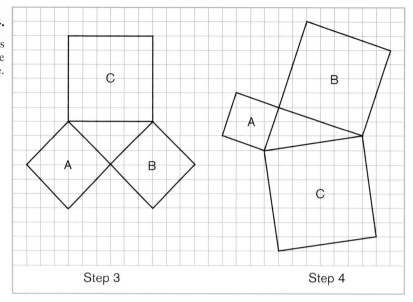

Step 3   Step 4

If you don't want to have your students decompose the square on the hypotenuse to find its area, you can have them count the squares, combining and accounting for partial squares. This can be a tedious process, however, with opportunities for error.

Another possibility is to have the students work the other way around—from side length to area instead of from area to side length. If you wish, your students can simply measure the side of the square by using a ruler calibrated in centimeters and millimeters, determining measures to the nearest tenth of a centimeter. (The figures on the activity page appear on centimeter grids for this purpose.) Then the students can square the side length to calculate the area of the square.

Select the method that suits your purposes, and let your students complete steps 2–4 on their own. These steps have the same form as step 1: the students find the areas and side lengths of all three squares on the sides of a right triangle, and then they speculate about the relationship among the areas of the squares. The right triangle in step 1 is a neat 3-4-5 triangle, but it is the only right triangle with all whole-number side lengths that the students encounter in steps 1–4. (For example, the length of the hypotenuse of the right triangle in step 2 is equal to $\sqrt{20}$ linear grid units. Students who decompose the square on the hypotenuse as shown in figure 6.3 and use a calculator to estimate $\sqrt{20}$ might give the length as approximately 4.472 units; students who are measuring with a ruler might estimate it as 4.4 or 4.5 centimeters.)

Steps 3 and 4 show triangles that are oriented in different ways on the grid (see fig. 6.4). In each case, however, the triangle's vertices continue to be at the intersections of grid lines, and as a result, the students can use the same method of decomposition as before. However, depending on the right triangle's orientation with respect to grid lines, the students may have to decompose two squares on the triangle's sides, or all three squares, instead of just one, as in steps 1 and 2 (see fig. 6.5).

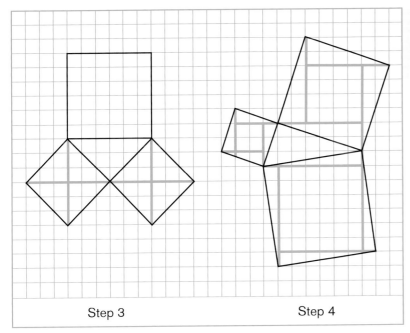

Step 3   Step 4

Moreover, as figure 6.5 illustrates, the students will observe in step 3 that they can decompose each of the squares on the legs of the right triangle into the four congruent triangles shown without creating an enclosed, smaller square. (Alternatively, they can decompose each of these squares into two congruent triangles.) The orientation of the legs of this right triangle with respect to the grid—each leg forms a 45-degree angle with the grid lines—eliminates the small square from the decomposition. Despite the slight variations in steps 3 and 4, however, partitioning continues to be a relatively easy way for the students to determine the areas of the squares constructed on the sides of the triangles, as figure 6.5 shows.

The students' work in steps 1–4 will probably quickly suggest to them—if they weren't already aware—that the area of the square on the hypotenuse is equal to the sum of the areas of the squares on the legs. Students whose understanding of proof is at level 1 will recognize the need to prove this idea, but on the basis of their work with a few specific examples, they may want to claim that the relationship that they have observed in the examples always holds, without being able to construct an argument that goes beyond an assertion such as, "I think it's true, because that's what happened in each case."

The activity continues, however, leading the students gently to level 2 by helping them recognize the insufficiency of their four examples to prove their conjecture. Step 5 invites them to test more examples—the natural impulse of those who have attained level 2. If they have access to the applet Squaring the Triangle (see fig. 6.6), they can use it in this step to find and check more varied examples.

The applet displays a right triangle *ABC* with a square on each of its three sides. The interactive Pythagorean construction appears on a grid and looks very much like the constructions that the students have worked with so far—especially in steps 1 and 2, where the legs of the right triangle appear on grid lines. The left side of the applet screen displays data on the triangle's side lengths and angles, as well as the

Fig. **6.6.**

The applet Squaring the Triangle lets
users adjust the lengths of the legs of a right
triangle and see the resulting Pythagorean
construction and its numerical data.

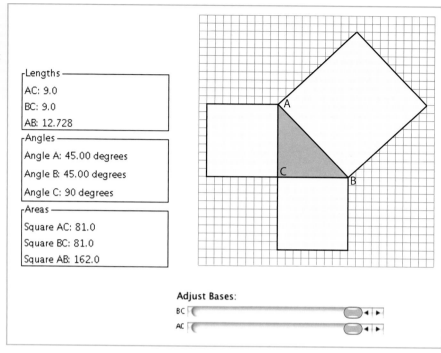

areas of the squares constructed on its sides. The students can use
two slides below the figure to adjust *AC* and *BC*—the lengths of the
triangle's legs. Any adjustments that they make register instantly in the
figure and the accompanying data.

Encourage your students to explore different conditions as they
work with the applet and complete the chart in step 5, entering data on
squares constructed on the sides of five new triangles. For example, they
can investigate cases in which leg *AC* of the right triangle is shorter than
leg *BC*, cases in which leg *BC* is shorter than leg *AC*, and cases in which
legs *AC* and *BC* are equal. In each instance, they should make observations about the relationship among the areas of the squares. Figure 6.7
shows the chart with sample values from the applet.

Fig. **6.7.**

The chart in step 5, with sample values
obtained from the applet Squaring
the Triangle

Lengths (linear grid units)			Areas (square grid units		
AC	*BC*	*AB*	*(AC)²*	*(BC)²*	*(AB)²*
5.0	4.0	6.403	25.0	16.0	41.0
3.0	8.0	8.544	9.0	64.0	73.0
2.0	8.0	8.246	4.0	64.0	68.0
9.0	1.0	9.055	81.0	1.0	82.0
5.0	5.0	7.071	25.0	25.0	50.0

If you are doing step 5 as a whole-class demonstration on one computer, or if you do not have a computer in your classroom, give your
students grid paper and have them make other right triangles, draw the

quares on the sides, and find the squares' areas and side lengths. They an fill in the chart in step 5 with the data from their own explorations. f you are demonstrating the applet on a single computer, you may be ble to recreate their examples and confirm their results. Step 6 wraps p this part of the investigation by asking studetns to make a clear statement of their conjecture on the basis of all the work that they have done o far.

Evaluate

One way to review the results of your students' work is to make a arge chart based on the one that the students themselves complete in tep 5 (see fig. 6.7). Use a sheet of chart paper, the board, or an overhead transparency for your chart, and display the solutions for the problems in steps 1–4 as well as a sampling of data on different triangles nd constructed squares in step 5.

Spend some time discussing how the students determined the length of the hypotenuse in each of these problems. Although many students will relate their reasoning to area of a square or a direct measurement of the length, some may already be applying the Pythagorean theorem nformally on the basis of "hearsay" or a few examples, without being ble to say for sure that the theorem is true. Students at this level need o continue to collect examples in order to solidify their conjecture; hus, it is important to make sure that the class verifies the areas of the quares in one fashion or another.

Be sure to talk about side lengths that are irrational numbers. Discuss each case to probe students' understanding of the particular estimates hat they made.

Call several students to the board to write the general statements hat they made in step 6 about the relationship among the areas of the quares. Let the class examine each statement, and discuss the need for clear and precise wording. Name the theorem that they have developed, nd ask them if they think that they could prove it *for any right triangle*.

Extend

Your students are now set to think about a general proof of the Pythagorean theorem. Their work with right triangles with squares constructed on the sides has led them to a conjecture, which they have ested with multiple examples. They have also probably begun to recognize that one successful example is much like another in its ability or nability to prove the conjecture. Moreover, they may have discovered hat amassing successful examples is something like sitting in a car that s spinning its wheels. Everything under the hood is functioning as well as anyone could wish, but the car is nevertheless churning in the same spot without making any headway. Is the conjecture true in every case? At this point, the students think that it is, but they would like to know for sure, and they are persuaded that having this certainty would be valuable. Yet, they probably do not know how to show that the conjecture is true in every case.

The students have gone from level 1 to level 2 to level 3 in their thinking about the Pythagorean theorem. At level 3, students are aware of the need for a generalized proof, and, although they are unable to construct a valid proof unaided, they can understand and follow a proof at an appropriate level of difficulty. Indeed, by using geometric and algebraic methods, students can explore a more general proof of the

Pythagorean theorem. The proof considered here depends on some knowledge of and skill with symbolic algebraic representations.

Give each student one or two sheets of centimeter grid paper as well as copies of the activity sheets "Can You Prove It?" and "Sets of Right Triangles for 'Can You Prove It?'" Again have the students work in pairs (or in small groups of up to four members) so that they can discuss their ideas as they work. Write a clear and complete statement of the theorem on the board: The sum of the areas of squares constructed on the legs of a right triangle is equal to the area of a square constructed on the hypotenuse.

In step 1, the students cut out three sets of four congruent right triangles. Leg *a* is longer than leg *b* in one set, equal to leg *b* in another set, and shorter than leg *b* in a third set. In step 2, the students work with the sets one at a time on grid paper, using each set of four triangles to make a square that has a side length of *a* + *b* and circumscribes a second square with a side length of *c* (see fig. 6.8). (The students assume but do not prove that the inscribed figure is a square. They could make the proof by considering the straight angle on the side of the larger square. This angle is formed by the acute angles of a right triangle together with an angle of the inscribed figure, so this angle must measure 90 degrees.) In step 3, the students draw the three sets of circumscribed and inscribed squares on grid paper, as they appear in figure 6.8, with the sides of the circumscribed squares on grid lines.

Fig. **6.8.**

A set of four congruent right triangles with legs of length *a* and *b* and a hypotenuse of length *c* can form a square that has a side length of *a* + *b* and circumscribes a square with a side length of *c*.

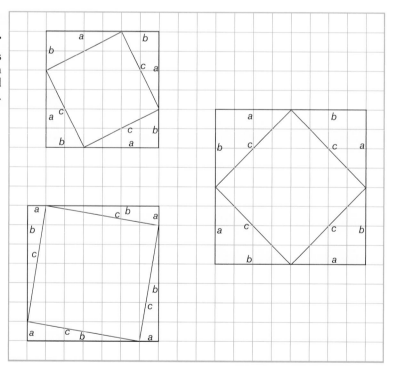

Steps 4, 5, and 6 all have the same form. For each square with an inscribed square, the students consider that they can express the area of the circumscribed square in two ways: (1) as the side length squared and (2) as the sum of the areas of the component regions. If they think of the area of the circumscribed square in the first way, they write the expression

$$(a + b)^2, \text{ or } a^2 + 2ab + b^2,$$

for its area. If, however, they think of the area of this square in the second way, they write the expression

$$4 \times \left(\frac{a \times b}{2} \right) + c^2, \text{ or } 2ab + c^2$$

for the area. In the case of each square with an inscribed square, the students consider that these expressions are equivalent:

$$a^2 + 2ab + b^2 = 2ab + c^2.$$

Thus, $a^2 + b^2 = c^2$. The sum of the areas of the squares on the legs of the triangle is equal to the area of the square on the hypotenuse.

Students in grades 6–8 may not have had sufficient experience in manipulating algebraic expressions to follow this proof with ease. The binomial expansion of $(a + b)^2$ may be particularly challenging to those who have not begun to study algebra formally. If your students have difficulty with these expressions and manipulations, approach the proof in a different way.

To do so, have your students draw on grid paper three new copies of the circumscribed (larger, or "outer") squares that they created with the sets of right triangles. After they have drawn these three squares with side length $a + b$, have them use their sets of cut-out triangles one at a time, again partitioning the respective areas of these squares, but in a new way, as shown in figure 6.9. This time, let them use each set of four triangles to form two a-by-b rectangles. Direct them to join the rectangles in each pair at a vertex in such a way that the long sides of the rectangles form a right angle. In each case, have them then locate this construction inside the corresponding larger square, as shown.

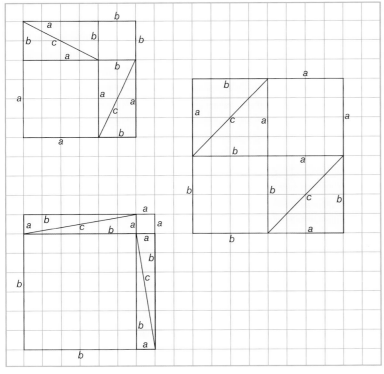

Fig. **6.9.**

A set of four congruent right triangles with legs of length a and b and a hypotenuse of length c can partition a square with side length $a + b$ in a different way to help prove that prove that $a^2 + b^2 = c^2$.

The students can compare their new drawing with their original drawing in each case. They can see that when they arranged the triangles in the original way, making the inscribed square with side length c, they naturally represented the area of the circumscribed square as $2ab + c^2$. However, when they arrange the triangles in the new way, as two $a \times b$ rectangles, they just as naturally represent the area of the same square as $a^2 + 2ab + b^2$. They can also see clearly that taking away the four congruent triangles from the original square leaves c^2, and taking them away from the new square leaves $a^2 + b^2$. But since the areas of the squares are equal, $a^2 + b^2 = c^2$.

This work also gives the students a geometric representation of the algebraic expansion of $(a + b)^2$ presented earlier. The new partitioning offers an area model for the multiplication of $(a + b)$ times $(a + b)$ that shows the product, $a^2 + 2ab + b^2$, very concretely. Use your students' understanding of the distributive property of multiplication over addition to help them understand this result. Point out that the product applies the distributive property twice:

$$(a+b) \times (a+b) = \big((a+b) \times a\big) + \big((a+b) \times b\big) \text{ (first application of the distributive property)}$$

$$= (a \times a) + (b \times a) + (a \times b) + (b \times b) \text{ (second application of the property)}$$

$$= a^2 + ba + ab + b^2$$

$$= a^2 + 2ab + b^2.$$

As a final extension of the activity, you might want to give your students an opportunity to experience the usefulness of the Pythagorean theorem. Present several right triangles, giving two side measurements and asking them to use the theorem to determine the missing side length. Have them find the length of a leg in some cases and the length of the hypotenuse in others. Emphasize that the certainty that the relationship holds in every case—certainty that the generalized proof gives—allows them to make these determinations. Stress the fact that this important theorem is indispensable to the study of geometry.

Conclusion

Being able to reason is essential to understanding mathematics. Students need regular opportunities to develop ideas and make, explore, and justify conjectures. In this way, they learn to expect that they will always be able to make sense of mathematics. Initially, students may rely more on inductive reasoning, but as they work with more sophisticated and abstract concepts, they can begin to reason deductively as well. Although no one expects students in the middle grades to construct formal proofs routinely or without assistance, you can begin to prepare your students for the rigorous thinking of proof by giving them opportunities to justify their reasoning and, whenever possible, to engage in activities that make them aware of the need for, and the methods of, formal proof.

PROBLEM SOLVING *and* REASONING

Looking Back and Looking Ahead

Given appropriate opportunities and guidance, every student has the potential to reason about mathematics:

> The key is to unlock the world of mathematics through a student's natural inclination to strive for purpose and meaning. Reasoning is fundamental to the knowing and doing of mathematics. Conjecturing and demonstrating the logical validity of conjectures are the essence of the creative act of *doing* mathematics. To give more students access to mathematics as a powerful way of making sense of the world, it is essential that an emphasis on reasoning pervade all mathematical activity. In order to become confident, self-reliant mathematical thinkers, students need to develop the capability to confront a mathematical problem, persevere in its solution, and evaluate and justify their results. (Rosenstein 1996, Standard 4—Reasoning)

When students reason, they recognize that they can understand and make sense of mathematics. They learn how to evaluate a problem, select an appropriate approach or strategy, and draw logical conclusions. After developing a solution, they reflect on it to determine whether or not it makes sense. They appreciate the pervasive use and power of reasoning in mathematics.

Students' experiences with reasoning and proof should span the entire mathematics curriculum at all grade levels. *Principles and Standards for School Mathematics* (NCTM 2000) contends that "reasoning and proof are not special activities reserved for special times or special topics in the curriculum but should be a natural, ongoing part of classroom discussions, no matter what is being studied" (p. 342). Students who are continually faced with "why" questions about the mathematics they are studying become adept at justifying their thinking and broadening their approaches to mathematics.

Such experiences are essential at the middle school level if students are to be successful in secondary mathematics classrooms. As they make the transition from the hands-on, concrete work of elementary school to the more formal, abstract work of secondary school, their abilities to reason about mathematics are expanding. They have had many inductive reasoning experiences, in which they have drawn conclusions on the basis of observations, and they are ready to take their first steps toward deductive reasoning experiences, in which they will arrive at conclusions by combining facts according to principles of logic.

In grades 6–8, students gradually come to understand that they cannot consider a conjecture to be true simply because many examples support it. However, they are learning that a conjecture is categorically false if just one example fails to support it. Thus, they discover the power of a *counterexample* to refute a proposition and the inefficacy of multiple examples to prove one.

In addition, middle school students are beginning to value the role of deductive proof in establishing results. Reasoned explanations, logical arguments, and proofs—in paragraph, two-column, and other formats—are tools that students themselves use with increasing frequency, or they watch with growing understanding as others use them. They may be exposed to indirect proof or experience the power of mathematical induction. The foundation for mathematical reasoning that middle school experiences build provides students with various tools for reasoning and argument that support their mathematical inquiry in other settings and equip them with reasoning skills that they can use in everyday situations. In grades 9–12, "the repertoire of proof techniques that students understand and use should expand" (NCTM 2000, p. 345).

The Learning Principle enunciated in *Principles and Standards* provides a valuable guideline for mathematics instruction at all levels: "Students must learn mathematics with understanding, actively building new knowledge from experience and prior knowledge" (NCTM 2000, p. 20). Success in secondary school mathematics depends on learners' skill in reasoning about and making sense of the mathematics that they do each day. The best opportunity that you can offer your middle school students is the chance to have these experiences every time they "do" mathematics. With sufficient guidance and practice, middle-grades students can move into secondary mathematics well prepared for more advanced reasoning, problem solving, and proof.

PROBLEM SOLVING *and* REASONING

Appendix

Blackline Masters and Solutions

Sum-thing about Consecutive Numbers

Name_____

Examine the three sums of consecutive integers below:

$$3 + 4 = 7 \qquad 2 + 3 + 4 = 9 \qquad 4 + 5 + 6 + 7 = 22$$

Note that 7 appears as the sum of two consecutive numbers, 9 appears as the sum of three consecutive numbers, and 22 appears as the sum of four consecutive numbers. Which counting numbers (1, 2, 3, …) from 1 to 35 can be the sums of consecutive counting numbers? Which cannot?

1. Find all the ways in which you can write the counting numbers from 1 to 35 as sums of two or more consecutive counting numbers.

1		2	
3		4	
5		6	
7		8	
9		10	
11		12	

(continued)

This activity is adapted from Mark Driscoll, *Fostering Algebraic Thinking: A Guide for Teachers, Grades 6–10*, Education Development Center (Portsmouth, N.H.: Heinemann, 1999), p. 79.

Navigating through Problem Solving and Reasoning in Grades 6

Sum-thing about Consecutive Numbers (continued)

Name______________________________

13		14	
15		16	
17		18	
19		20	
21		22	
23		24	
25		26	
27		28	
29		30	

(continued)

Name_____

31		32	
33		34	
35			

2. Have you discovered any patterns in sums of the consecutive addends? On "Recording Sheet for 'Sum-thing about Consecutive Numbering'" organize your data according to the number of addends in each sum. Write out your findings on a sheet of chart paper, and be prepared to share your ideas.

3. Without doing any calculations, predict whether each of the following numbers is the sum of two, three, four, or more consecutive counting numbers. Explain your predictions in each case.

45

57

62

75

80

Navigating through Problem Solving and Reasoning in Grades 6

Name_____

4. Do your discoveries in step 2 suggest any shortcuts that you could use for determining which of the following numbers are the sums of two or more consecutive addends? Explain your shortcuts and tell how to use them to write each number below as the sum or sums of consecutive numbers.

45

57

62

75

80

Recording Sheet for "Sum-thing about Consecutive Numbers"

In step 1 of "Sum-thing about Consecutive Numbers," you discovered which numbers from 1 to 35 are the sums of consecutive counting numbers. Below, display your strings of addends again, this time sorting them according to how many consecutive addends are in a string for a particular sum.

Sum	Number of Consecutive Addends					
	2 Addends	3 Addends	4 Addends	5 Addends	6 Addends	7 Addends
1						
2						
3						
4						
5						
6						
7						
8						
9						
10						
11						
12						
13						
14						

Name _____

Sum	2 Addends	3 Addends	4 Addends	5 Addends	6 Addends
15					
16					
17					
18					
19					
20					
21					
22					
23					
24					
25					
26					
27					
28					
29					
30					
31					
32					
33					
34					
35					

Looking for the Least

Name_____

Reasoning about Factors and Multiples—Part 1

1. Suppose that grocery stores always sell hot dog rolls in packages of 8 and hot dogs in packages of 10. You are planning the food for a field trip with your class. Everyone wants to eat two hot dogs.

    a. What is the smallest number of packages of hot dogs and the smallest number of packages of rolls that you would need to buy to have equal numbers of hot dogs and rolls?

    b. How many students will you be able to feed with this many packages of hot dogs and this many packages of rolls?

    c. Explain your reasoning.

    d. The count for the field trip is final. You now know that 24 students, 4 parents, your teacher, and the bus driver are all going to be eating lunch on the field trip. You want to buy the fewest packages of hot dogs and rolls to ensure that everyone will have two hot dogs. How many packages of hot dogs and how many packages of rolls do you need to buy?

    e. Explain your reasoning.

2. The light at the top of one lighthouse rotates 360 degrees in 15 minutes. The light at the top of another lighthouse makes a complete rotation of 360 degrees in 20 minutes.

    a. If the lights on the two lighthouses are synchronized to begin a revolution in the same position at 9:00 a.m., how much time will elapse before they are synchronized again?

    b. Explain your reasoning.

Looking for the Least (continued)

Name_____

Children need about one more hour of sleep each night than teenagers do, and teenagers, in turn, need about one more hour of sleep each night than adults do. Research studies show that in a 24-hour day—

- children need about 10 hours of sleep;
- teenagers need about 9 hours of sleep;
- adults need about 8 hours of sleep.

(See, for example, http://www.stanford.edu/~dement/adolescent.html.) Suppose that a child does sleep 10 hours, a teenager sleeps 9 hours, and an adult sleeps 8 hours, each night.

a. What is the smallest number of nights that the child must sleep and the smallest number of nights that the teenager must sleep for both of them to have slept the same number of hours?

The child must sleep _____ nights.

The teenager must sleep _____ nights.

Each of these numbers of nights is equal to _____ hours of sleep.

b. Explain your reasoning.

c. What are the smallest numbers of nights that the child, the teenager, and an adult must sleep for all of them to have slept the same number of hours?

The child must sleep _____ nights.

The teenager must sleep _____ nights.

The adult must sleep _____ nights.

Each of these numbers of nights is equal to _____ hours.

d. Explain your reasoning.

In-Venn-stigating Factors and Multiples

Name_____

Reasoning about Factors and Multiples—Part 2

1. *a.* In the appropriate regions in the Venn diagram below, enter all the multiples of 3 and 8 that are greater than 0 and less than 100.

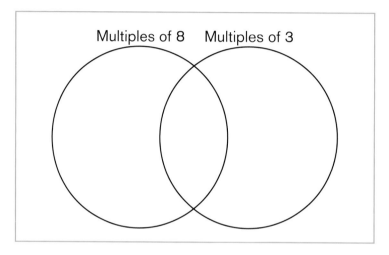

b. Explain how you can use the completed diagram to find common multiples of 3 and 8 that are great than 0 and less than 100.

2. *a.* In the appropriate regions in the Venn diagram below, enter all the multiples of 2 and 7 that are greater than 0 and less than 30.

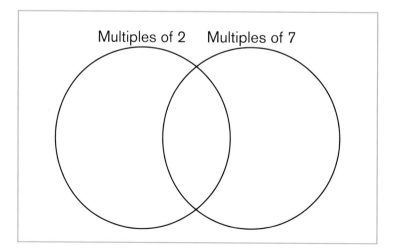

Name_____

b. Explain how you can use the completed diagram to find common multiples of 2 and 7 that are greater than 0 and less than 30.

c. List five more numbers that would be in the intersection if numbers greater than 30 were allowed.

3. Describe the numbers in the intersections of the Venn diagrams in steps 1 and 2. Why are these numbers in the intersections?

4. *a.* In the appropriate regions in the Venn diagram below, enter all the multiples of 5 and 15 that are greater than 0 and less than 70.

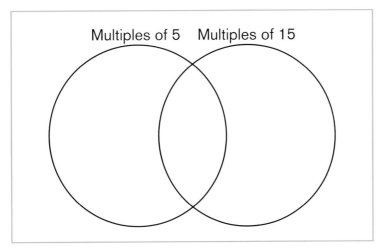

Multiples of 5   Multiples of 15

b. Explain how you can use the completed diagram to find common multiples of 5 and 15 that are greater than 0 and less than 70.

Name_____

5. *a.* In the appropriate regions in the Venn diagram below, enter all the multiples of 4 and 8 that are greater than 0 and less than 50.

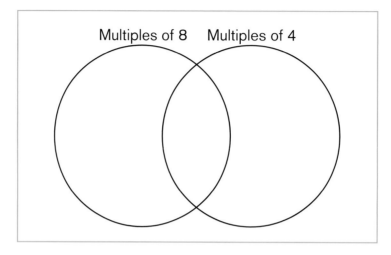

Multiples of 8   Multiples of 4

b. Explain how you can use the completed diagram to find common multiples of 4 and 8 that are grea than 0 and less than 50.

c. List five more numbers that would be in the intersection if numbers greater than 50 were allowed.

6. *a.* Describe the numbers in the intersections of the Venn diagrams in problems 4 and 5. Why are thes numbers in the intersections?

b. How are the pairs of numbers that you investigated in steps 4 and 5 different from the pairs of numbers that you investigated in steps 1 and 2? *Hint:* Consider the relationship between the numbe in each pair. How is this relationship shown in the distribution of the elements in the regions of the diagram?

c. Can you give another pair of numbers whose multiples would produce a Venn diagram like those that you completed in steps 1 and 2? _____ Make a Venn diagram for your pair of numbers.

d. Can you give another pair of numbers whose multiples would produce a Venn diagram like those that that you completed in steps 4 and 5? _____ Make a Venn diagram for your pair of numbers.

7. *a.* In the appropriate regions in the Venn diagram below, enter all the multiples of 8 and 12 that are greater than 0 and less than 100.

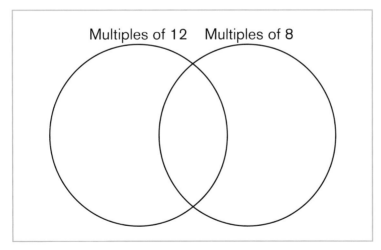

Multiples of 12   Multiples of 8

b. Explain how you can use the completed diagram to find common multiples of 8 and 12 that are greater than 0 and less than 100.

Name_____

c. Describe how you can use the information in the completed diagram to find the least common multiple (LCM) of 12 and 8. What is their LCM?

8. *a.* In the appropriate regions in the Venn diagram below, enter all the multiples of 14 and 21 that are greater than 0 and less than 170.

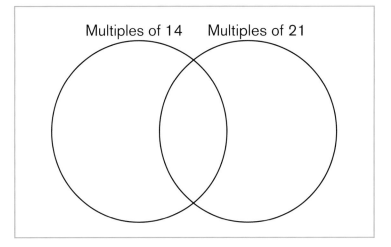

b. Explain how you can use the completed diagram to find common multiples of 14 and 21 that are greater than 0 and less than 170.

c. Describe how you can use the information in the completed diagram to find the least common multiple of 14 and 21. What is the LCM?

9. *a.* Describe the numbers in the intersections of the Venn diagrams in steps 7 and 8. Why are these numbers in the intersections?

Name_____

b. How are the pairs of numbers that you investigated in steps 7 and 8 different from the pairs of numbers that you investigated in steps 4 and 5?

c. Can you give another pair of numbers whose multiples would produce a Venn diagram like those that you completed in steps 7 and 8? _____ Make a Venn diagram for your pair of numbers.

a. Make a two-circle Venn diagram. Complete the diagram by entering the *factors* of 8 and 12 in the appropriate regions. Be sure to label your Venn diagram.

b. Explain how you can use the information in the completed diagram to find the greatest common factor (GCF) of 8 and 12.

Name_____

11. *a.* Make and label another two-circle Venn diagram in which you enter the factors of 14 and 21 in the appropriate regions.

b. Explain how you can use the information in the completed diagram to find the greatest common factor (GCF) of 14 and 21.

12. *a.* Make and label a two-circle Venn diagram in which you enter the *prime factors* of 8 and 12 in the appropriate regions.

b. Look back at steps 7 and 10. What are the GCF and the LCM of 8 and 12? Now look at your Venn diagram showing prime factors of 8 and 12. Explain how you can use the information in this diagram to find the GCF and LCM of 8 and 12.

Name_____

13. *a.* Make and label a two-circle Venn diagram in which you enter the *prime factors* of 14 and 21 in the appropriate regions.

b. Look back at steps 8 and 11. What are the GCF and the LCM of 14 and 21? Now look at your Venn diagram showing prime factors of 14 and of 21. Explain how you can use the information in this diagram to find the GCF and LCM of 14 and 21.

What Are the Relationships?

Name_____

Reasoning about Factors and Multiples–Part 3

1. Choose pairs of natural numbers (counting numbers) *A* and *B* to enter in the last six rows of the table below. Complete the table by filling in the factors of *A* and *B*, their GCF, prime factorizations, and LCM

A	*B*	Factors of *A*	Factors of *B*	GCF of *A* and *B*	Prime Factorization of *A*	Prime Factorization of *B*	LCM of *A* and *B*
4	6	1, 2, 4	1, 2, 3, 6	2	2×2	2×3	12
6	5	1, 2, 3, 6	1, 5	1	2×3	5	30
4	8	1, 2, 4	1, 2, 4, 8	4	2×2	$2 \times 2 \times 2$	8
8	28	1, 2, 4, 8	1, 2, 4, 7, 14, 28	4	$2 \times 2 \times 2$	$2 \times 2 \times 7$	56
6	23	1, 2, 3, 6	1, 23	1	2×3	23	138

2. Examine each pair of numbers in the table and its corresponding GCF and LCM.

   a. Can you find any relationship between the numbers in a pair and their GCF and LCM? Write any conjecture that you make.

Name_____

b. Do you think your conjecture is true? In other words, do you think that the relationship that you have identified always holds?

c. If you think your conjecture is true, explain why you think the relationship holds. *Hint:* Finding the relationships in steps 3 and 4 below may help you construct your explanation.

3. Can you find a relationship between the factors of two numbers and their GCF? Explain.

4. Can you find a relationship between the factors in the numbers, prime factorizations and their LCM? Between the factors in the prime factorizations and the GCF? Explain.

Parallel(ogram) Universe

Name_____

A *rectangle* is a quadrilateral whose sides form four right angles. You know how to find the area of a rectangle. A *parallelogram* is a quadrilateral whose opposite sides are parallel. Can you use your knowledge of the area of a rectangle to find the area of a parallelogram?

1. Draw on the parallelograms below and show any calculations that you need to make to determine the area of each parallelogram.

   a. Parallelogram 1

   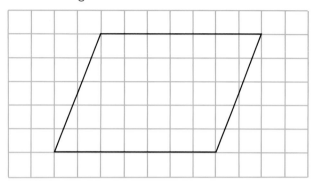

   b. Parallelogram 2

   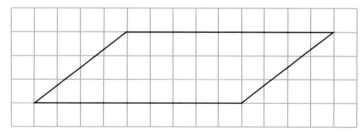

2. How did you determine the area of each of the parallelograms in step 1?

Name_____

3. Explain the process that you used clearly enough that a fifth grader who knows how to calculate the area of a rectangle can find the area of *any* parallelogram.

4. On the basis of your work, state a formula for the area of a parallelogram.

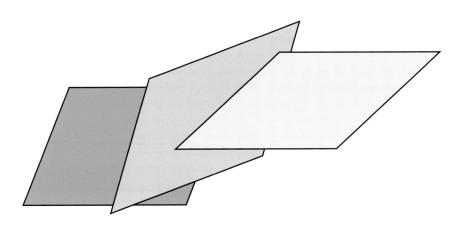

Moving into Triangular Territory

Name_____

In completing the activity sheet "Parallel(ogram) Universe," you used what you knew about determining the area of a rectangle to help you derive a formula for the area of a parallelogram. A *triangle* is a three-sided polygon. Can you use your earlier work with four-sided polygons—specifically, rectangles and parallelograms—to discover the formula for the area of a triangle?

1. Draw on the grids below and show any calculations that you need to make to determine the area of each triangle.

   a. Triangle 1

   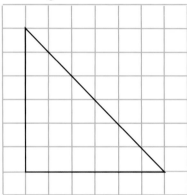

   b. Triangle 2

   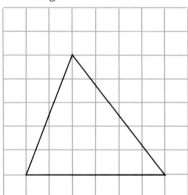

   c. Triangle 3

   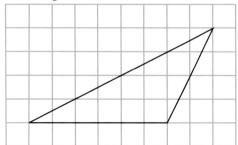

Name_____

2. How did you determine the area of each of the triangles in step 1?

3. Explain the process that you used in step 1 clearly enough that anyone who knows how to calculate the area of a rectangle and a parallelogram can find the area of any triangle.

4. On the basis of your work, state a formula for the area of a triangle.

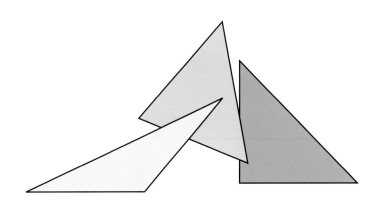

How Much Area Does a Trapezoid Trap?

Name_____

In completing the activity sheet "Parallel(ogram) Universe," you used what you knew about the area of a rectangle to discover a formula for the area of a parallelogram. Then, in "Moving into Triangular Territory," yo̶u̶ used your earlier work with these quadrilaterals to help you discover the formula for the area of a triangle.

A *trapezoid* is a quadrilateral that has exactly one pair of parallel sides. Can you use your work on both previous tasks to discover the formula for the area of a trapezoid?

1. Parts (*a*) and (*b*) show different trapezoids. Can you determine the area of each trapezoid *by at least two different methods*? Draw on the figures and show any calculations that you need to make to determine the area of each trapezoid.

a. Trapezoid 1

Method 1

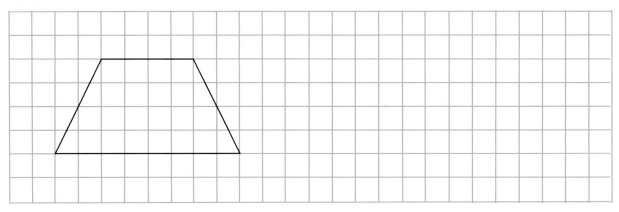

Method 2

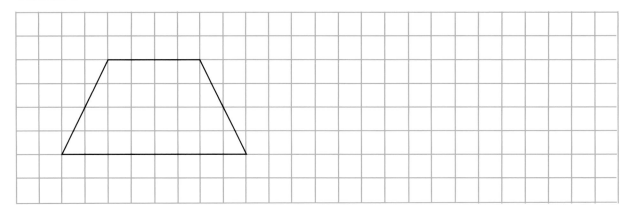

Name_____

Method 3 (optional)

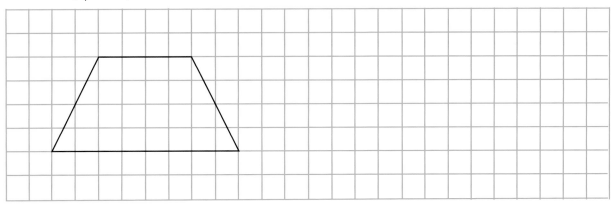

b. Trapezoid 2

Method 1

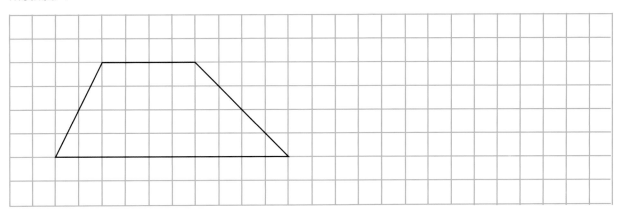

Method 2

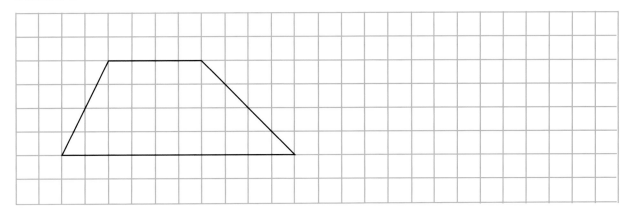

Name_____

Method 3 (optional)

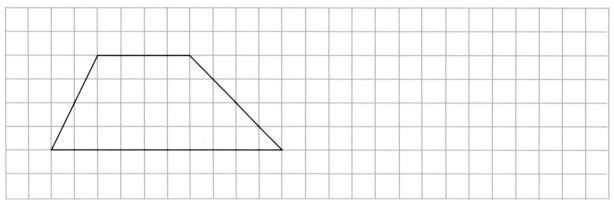

2. Could you use each method that you used in step 1 to determine the area of any trapezoid? Explain.

3. Using a method that you believe works for any trapezoid, calculate the area of each of the trapezoids below.

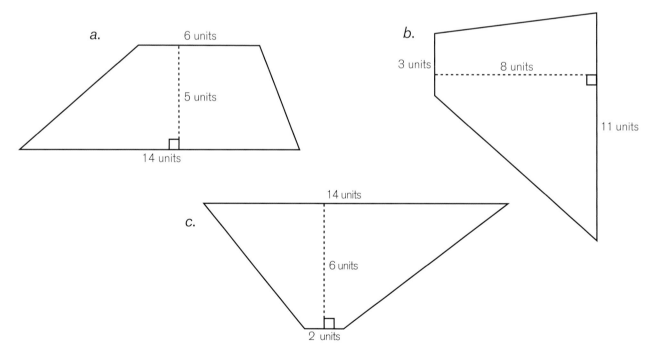

4. On the basis of your work, can you state a formula for the area of any trapezoid?

Carryover to Kites

Name_____

A *kite* is a quadrilateral with two pairs of congruent adjacent sides and exactly one line of symmetry. Be-
[ca]use a kite has only one line of symmetry, all four of its sides cannot be congruent. The quadrilateral on the
[lef]t below is a kite. Find its line of symmetry. The quadrilateral on the right has four congruent sides. Like a
[kit]e, it has two pairs of congruent adjacent sides—but it has two lines of symmetry instead of just one. Find
[its] lines of symmetry. This quadrilateral is a rhombus.

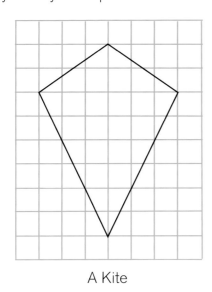

A Kite

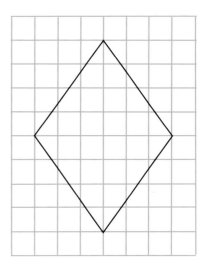

A Rhombus

In your earlier work, you discovered formulas for the areas of a parallelogram, a triangle, and a trapezoid.
[C]an you use any of this work to discover a formula for the area of a kite?

1. Parts (*a*) and (*b*) show different kites. Can you determine the area of each kite *by two different
   methods*? Draw on the figures and show any calculations that you need to make to determine the
   area of each kite.

   a. Kite 1

Method 1	Method 2

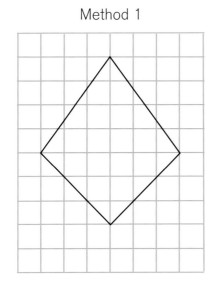 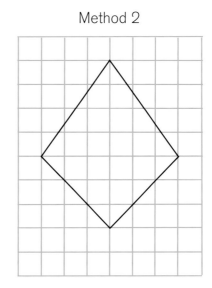

Name_____

b. Kite 2

Method 1   Method 2

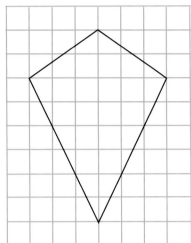   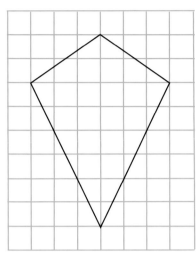

2. Could you use each method that you used in step 1 to determine the area of any kite? Explain.

3. Using a method that you believe works for any kite, calculate the area of each of the kites below.

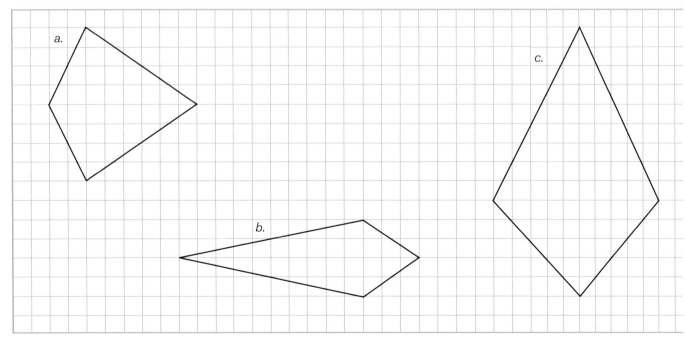

4. On the basis of your work, can you state a formula for the area of a kite?

Stepping Up to Stairs

Name_____

A "stair" is not a quadrilateral—or any other formally named mathematical shape. In this activity, a stair is
closed figure composed of several adjacent rectangles. All the bases of the rectangles are equal and
t on the same line, and the absolute difference between the heights of any two adjacent rectangles is a
nstant. Thus, the rectangles can be arranged as a two-dimensional "staircase," with steps ascending from
e "shortest" rectangle to the "tallest." Two stairs appear below.

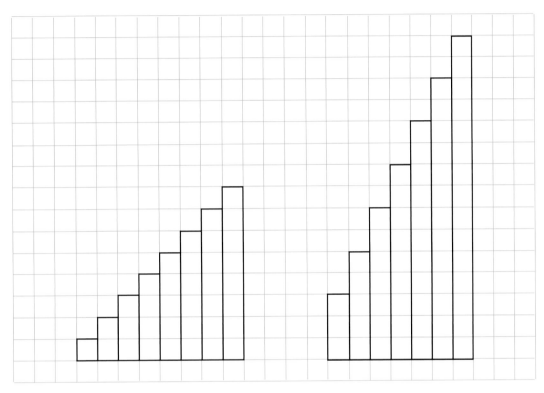

A simple way of determining the area of a stair is to calculate the area of each individual rectangle and
d sum of all the areas. However, as you can imagine, this process might be time-consuming and inefficient.

1. Using techniques and area formulas that you have developed in your earlier work, can you find an easier
   way to determine the areas of the stairs shown above?

2. On the basis of your work, can you state a formula for the area of a stair?

Diagonal Discoveries

Name_____

1. Cut out the five rectangles on the "Rectangles for 'Diagonal Discoveries'" sheets.

2. In the table below, record each rectangle's length, width, perimeter, and area. Be sure to include the units for each measurement. Also find and record the ratio of each rectangle's width to its length.

Rectangle	Length (cm)	Width (cm)	Perimeter (cm)	Area (cm²)	Width/Length
Rectangle 1					
Rectangle 2					
Rectangle 3					
Rectangle 4					
Rectangle 5					

3. Inspect the data that you collected in step 2. Can you tell from the data whether any rectangle has a special relationship with any other rectangle or rectangles in the set?

4. *a*. On each of the five rectangles, draw the diagonal from vertex *A* to vertex *C*.

   b. Position the rectangles so that *A* is at the bottom left and *C* is at the top right.

   c. Stack the five rectangles from largest to smallest, aligning all vertices *A*, sides *AB*, and sides *AD*, so that the five rectangles are "nested," as illustrated below.

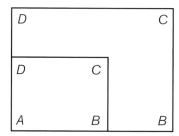

Navigating through Problem Solving and Reasoning in Grades 6–

Name_____

5. What relationship do you find among the diagonals of some of the rectangles?

6. Can you formulate a generalization that states this relationship?

7. Can you create a mathematical argument to support the generalization that you made in step 6?

Rectangles for "Diagonal Discoveries"

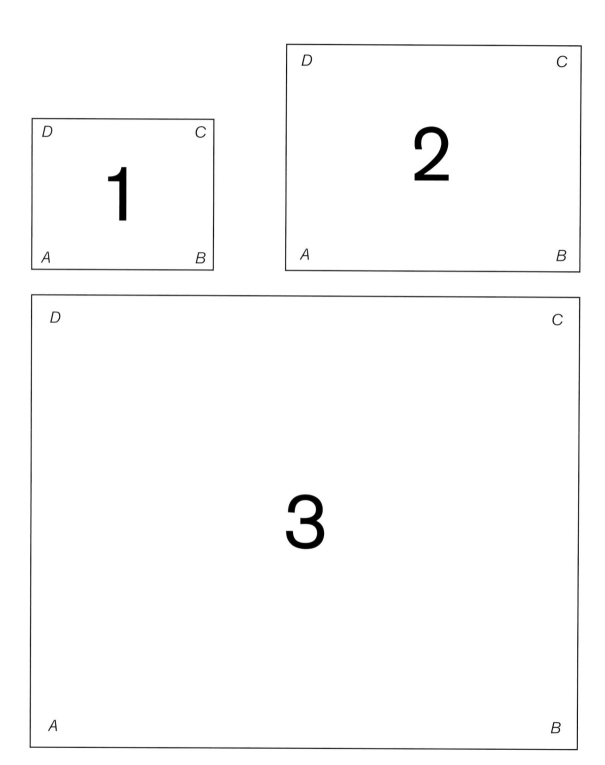

Navigating through Problem Solving and Reasoning in Grades 6–

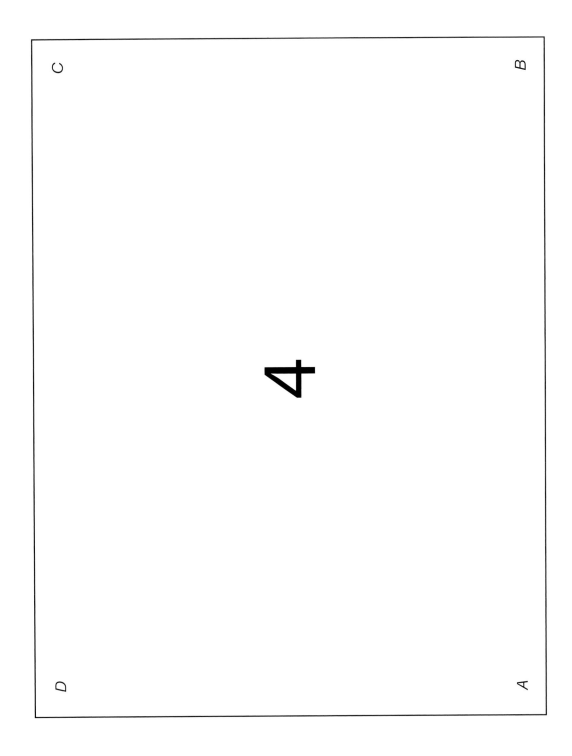

Popular People

Name_____

Many factors determine a celebrity's popularity. The table below displays data on five factors for twenty celebrities: (1) total earnings, (2) the number of hits on a celebrity's Web site, (3) the number of newspaper articles in which a celebrity's name appeared, (4) the number of magazine cover stories on a celebrity, and (5) the number of TV or radio reports about a celebrity.

Name	Earnings (in millions of dollars)	Number of Web Site Hits	Number of Newspaper Articles	Number of Magazine Cover Stories	Number of TV or Radio Reports
Jennifer Aniston	$18.5	1,110,000	8,176	5.5	201
Jerry Bruckheimer	66	300,000	3,570	0	40
Bill Clinton	6	5,760,000	80,075	1.5	5,629
Tom Cruise	31	2,650,000	15,784	4	343
Johnny Depp	37	1,930,000	11,682	5.5	169
Cameron Diaz	13	1,820,000	6,546	0.5	79
The Eagles	45	719,000	1,443	0	37
Mel Gibson	185	2,030,000	12,692	2.25	355
Michael Jordan	33	1,130,000	17,591	0	225
Jennifer Lopez	17	3,060,000	15,693	5.75	328
George Lucas	290	3,860,000	8,725	1	154
Madonna	50	3,100,000	13,488	0	286
Prince	49.7	1,190,000	830	0	6
Julia Roberts	8	1,870,000	11,257	9	249
J.K. Rowling	59.1	1,350,000	4,343	0	53
Steven Spielberg	80	2,110,000	11,971	0	253
Shania Twain	34	975,000	5,766	1	79
Denzel Washington	30	1,340,000	7,217	0	122
Oprah Winfrey	225	5,710,000	21,671	2.5	1,088
Tiger Woods	87	2, 050,00	48,467	1	11

The data are from *Forbes* magazine (http://www.forbes.com/lists/2005/53/Rank_1.html) for the time period from June 2004 to June 2005.

Note: Decimal fractions appear in the data on magazine cover stories. A fraction of a cover story reflects "sharing" of the story with others—for example, 0.5 represents a story shared with one other person, 0.25 represents a story shared with three others, and so on.

Who Is Most Popular?

Name_____

1. Use the data on the sheet "Popular People" to identify the three celebrities that you think are the most popular from the twenty people listed in the table.

     Most popular _____

     Second most popular _____

     Third most popular _____

2. Explain the method you used to determine the three most popular celebrities, detailing why you chose this method and why you think it works, as well as any mathematics that you used in your analysis.

How Do the Dollars Stack Up?

Name_____

The dot plot in graph 1 below shows the earnings of the twenty celebrities identified in the table on the tivity sheet "Popular People." Note that three fields (white, gray, white) differentiate the data into three oups—earnings less than $50 million, earnings equal to or greater than $50 million but less than $100 llion, and earnings of $100 million or more.

Graph 1

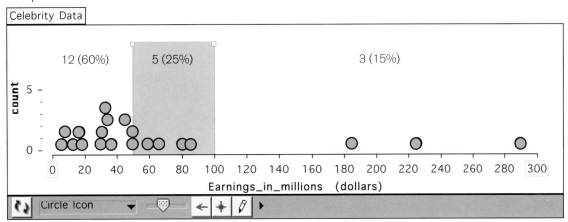

Mean = $68.2 million; median = $41 million

In graph 2 below, outliers are filtered—that is, hidden, or removed from consideration. Two fields (white and ay) differentiate the data into two groups—earnings less than $50 million and earnings equal to or greater an $50 million.

Graph 2

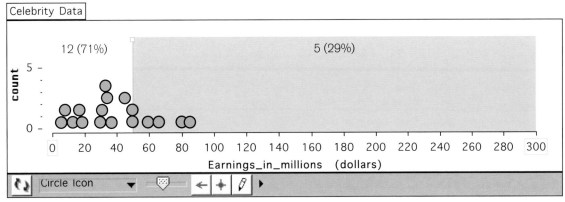

Earnings_in_millions < 100

Mean = $39.1 million; median = $34 million

Name_____

1. On the basis of graph 1, describe the variability in the celebrities' earnings.

2. Referring to graph 2, describe what happens to the measures of center for the data when outliers are filtered. Explain why these changes occur.

Word Spreads

Name_____

Below is a scatterplot to help you consider the relative popularity of the celebrities listed in the table on e activity sheet "Popular People" in terms of the publicity that they received in print.

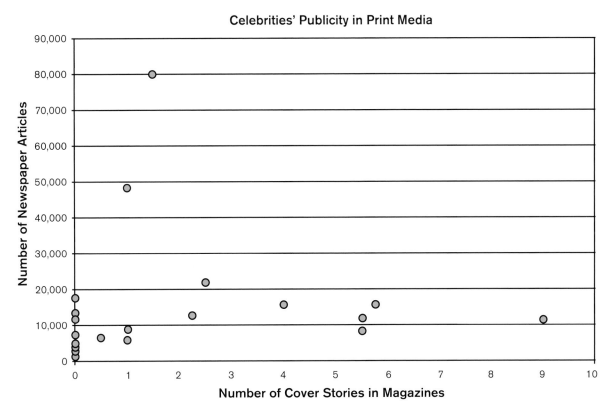

Celebrities' Publicity in Print Media

The graph shows the number of newspaper articles against the number of magazine cover stories for ch celebrity in the set of twenty. Remember that the data count newspaper stories in which celebrities' mes appeared even just once. Also recall that decimal fractions in data on magazine cover stories reflect pries that celebrities "shared" with others—for example, 0.5 represents a story that a celebrity shared with le other person, 0.25 represents a story that a celebrity shared with three others, and so on.

Investigate the publicity that celebrities in the sample received by looking at the distribution of the data.

1. Describe the variability in the numbers of cover stories about the celebrities in magazines.

Name_____

2. Describe the variability in number of newspaper articles reported for the celebrities.

3. What point or points would you regard as outliers in this graph? Would "filtering out" these points help you think about the distribution and interpret the data?

4. Refer to the table in "Popular People" to establish the identity of the celebrity who you think received the most publicity in print? _____

Change the Data–
Change the Stats?

Name_____

You have just worked with the activity sheet "TV Watching," which provides data on the hours in a week at students in two eighth-grade classes spent watching TV. Below is a bar graph of those data in the interval from 0 to 15 hours. The graph has been created electronically with a computer applet named op It! The screen image displays data in this interval for 43 students.

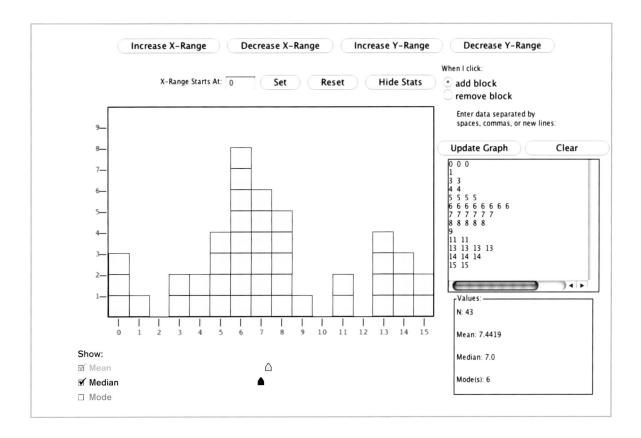

If you have access to Plop It! enter values and display them so that the image on your screen reproduces e image above. (If you are unable to work with Plop It! create the graph on chart paper, using sticky notes represent the data.)

How do the data in the set affect the locations of the median and the mean (both of which are marked by rows at the bottom of the screen in Plop It!)? Experiment to find out by making changes in selected data ues. As you change values for the data, pay attention to the locations and values of the median and the ean. Do they change as well?

Before you make a change in the data, try to predict what you think will happen to the measures of center, d why.

Name_____

1. Suppose that the two students who recorded 11 hours spent watching TV actually watched 13 hours apiece during the week.

   a. Predict what would happen to the median and the mean as a result of entering these changes.

   b. Change the data by adjusting these two students' hours upward to 13 hours. Record the results.

2. Leaving the data as adjusted in step 1, suppose that three of the five students who originally recorded hours of TV apiece meant instead to record 5 hours of TV watching.

   a. Predict what would happen to the median and the mean as a result of entering these changes.

   b. Change the data by adjusting three students' hours of TV watching down from 8 to 5 hours. Record the results.

3. Explain why the median changed when you made the changes in step 2.

4. Clear this distribution of data from the screen (or remove the sticky notes from your chart paper). Ente values to make a new graph to display a data set that consists of the numbers of hours that 12 people spent watching TV in a week. Let your set have the following summary statistics: its mean is 5 hours, its median is 5 hours, and its maximum value—the greatest amount of time that anyone spent watching TV—is 8 hours.

   a. Change five different data values in your new set in such a way that the new mean is greater than 5 hours while the median continues to be 5 hours, and the maximum continues to be 8 hours. Explain the reasoning behind your changes.

   b. Look at the distribution of your new data. What will happen to the median and the mean if you make one of the two smallest data values greater than the median? Explain your answer.

Name_____

c. Look at the altered distribution in step 4(b). What will happen to the median and the mean if you change the maximum data value to 15 hours? Explain your answer.

5. Look at the distribution of TV watching times on the graph below.

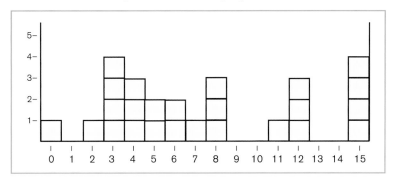

a. Without doing any computations, estimate the location of the median.

Explain your reasoning.

b. Do you think the mean be greater than, the same as, or less than the median?

Explain your reasoning.

Growing in Front of Your Eyes

Name_____

The first three stages in a growing pattern of black and white square tiles appear below.

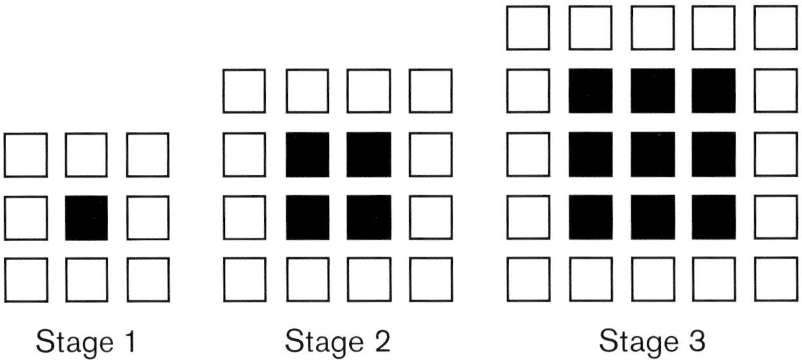

Stage 1	Stage 2	Stage 3

1. In the space below, draw and label stage 4.

   a. How many tiles (black and white tiles together) does stage 4 have in all? _____
   How did you know to draw this number of tiles?

   b. How many black tiles does stage 4 have? _____ How did you know to draw this number of black tiles?

Name_____

c. How many white tiles does stage 4 have? _____ How did you know to draw this
number of white tiles?

2. Without drawing this time, imagine that you are looking at stage 6 in the pattern.

   a. How many black tiles does stage 6 have? _____ Explain why you think so.

   b. How many white tiles does stage 6 have? _____ Explain why you think so.

   c. How many tiles (black and white together) does stage 6 have in all? _____
   Explain how you know.

3. Use the table on the next page to explore how are the numbers of black tiles, white tiles, and total tiles
   grow from one stage to the next.

   a. Talk with your partner(s) about combinations of words and numbers that you can use to describe
   what you see in the pattern from stage to stage. Enter these descriptions in columns 2–4. Note that
   different ways of "seeing" the numbers may be equally valid. For now, leave the table blank below the
   row for stage 6.

Name_____

Stage	Number of Black Tiles—Describe What You See	Number of White Tiles—Describe What You See	Total Number of Tiles—Describe What You See
1	A 1 × 1 square made up of 1 tile		
2	A 2 × 2 square made up of 4 tiles		
3			
4			
5			
6			
⋮	⋮	⋮	⋮
n			

Name_____

b. Now consider the last row in the table. Suppose that *n* in "stage *n*" stands for any natural number (1, 2, 3, …). In column 2 in the last row, write a mathematical expression using *n* that will give you the number of black tiles at stage *n*.

c. Write a mathematical expression using *n* that will give you the number of white tiles at stage *n*. Enter your expression in column 3 in the last row of the table.

d. Write a mathematical expression using *n* that will give you the total number of tiles at stage *n*. Enter your expression in column 4 in the last row of the table.

4. Imagine that you are looking at stage 17 in the pattern. Do you think that stage 17 has an odd number or an even number of black tiles? _____ Why do you think so?

5. *a.* Suppose that a particular stage in the pattern has 44 white tiles. What is *n*, the number of the stage?

b. How many blacks tiles does this stage have?

c. How many tiles in all (black and white tiles together) does this stage have?

Name_____

6. *a*. Is it possible for any stage in the pattern to have a total of 52 tiles (black and white tiles together)?

_____ Why, or why not?

b. Is it possible for any stage to have 52 black tiles? _____ Why, or why not?

c. Is it possible for any stage to have 52 white tiles? _____ Why, or why not?

Change the Pattern—
Change the Growth?

Name_____

The first three stages in a growing pattern of black and white tiles appear below. This pattern is different
·m the one on the activity sheet "Growing in Front of Your Eyes." Note that the stages in this pattern are
·issing" tiles that would make them square, as in the other pattern.

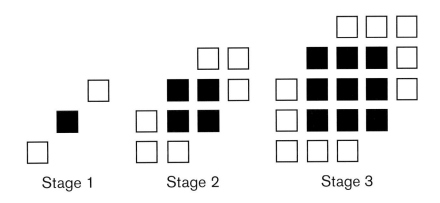

Stage 1   Stage 2   Stage 3

1. In the space below, draw and label stage 4.

a. How many black tiles does stage 4 have? _____ How did you know to draw this
number of black tiles?

Name_____

b. How many white tiles does stage 4 have? _____ How did you know to draw this number of white tiles?

c. How many tiles does stage 4 have in all (black and white tiles together)? _____ How do you know?

2. Imagine that you are looking at stage 5.

a. How many black tiles does stage 5 have? _____ Why do you think so?

b. How many white tiles does stage 5 have? _____ Why do you think so?

c. How many tiles (black and white together) does stage 5 have in all? _____ How do you know?

3. Use the table on the next page to explore how are the numbers of black tiles, white tiles, and total tile. grow from one stage to the next.

a. Talk with your partner(s) about combinations of words and numbers that you can use to describe what you see in the pattern from stage to stage. Enter these descriptions in columns 2–4. Note that different ways of "seeing" the numbers may be equally valid. For now, leave the table blank below the row for stage 6.

Name_____

Stage	Number of Black Tiles—Describe What You See	Number of White Tiles—Describe What You See	Total Number of Tiles—Describe What You See
1	A 1 × 1 square made up of 1 tile		
2			
3			
4			
5			
6			
⋮	⋮	⋮	⋮
n			

Name_____

b. Now consider the last row in the table. Suppose that n in "stage n" stands for any natural number (1, 2, 3, …). In column 2 in the last row, write a mathematical expression using n that will give you the number of black tiles at stage n.

c. Write a mathematical expression using n that will give you the number of white tiles at stage n. Enter your expression in column 3 in the last row of the table.

d. Write a mathematical expression using n that will give you the total number of tiles at stage n. Enter your expression in column 4 in the last row of the table.

4. Imagine that you are looking at stage 20 in the pattern. Do you think that this stage has an odd

number or an even number of black tiles? _____ Why do you think so?

5. a. Suppose that some stage in the pattern has 46 white tiles. What is the number of the stage?

b. How many blacks tiles does this stage have?

c. How many tiles does this stage have in all (black and white tiles together)?

Name_____

6. *a.* Is it possible for any stage to have a total of 115 tiles (black and white together)? _____

   Why, or why not?

   b. Is it possible for any stage to have 115 black tiles? _____ Why, or why not?

   c. Is it possible for any stage to have 115 white tiles? _____ Why, or why not?

What Changes?
What Stays the Same?

Name_____

1. The illustration below shows stage 2 in a growing pattern of square tiles in the shape of the capital letter H.

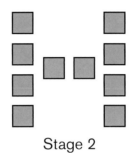

Stage 1   Stage 2   Stage 3

a. Suppose that stage 1 has 9 tiles and stage 3 has 11 tiles. What part of the H is changing from stage

to stage? _____ How do you know?

b. Draw stage 1 and stage 3 over the respective labels above.

c. What stays the same from stage to stage in the pattern?

d. Suppose that you are looking at stage 10 in the pattern. How many tiles does this stage have?

e. Why do you think your answer in (d) is correct?

Name_____

2. The illustration below shows stage 2 in a growing pattern of square tiles in the shape of the capital letter S.

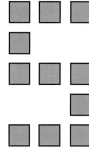

Stage 1   Stage 2   Stage 3

a. Suppose that stage 1 has 8 tiles and stage 3 has 14 tiles. What part of the S is changing from stage to stage? _____ How do you know?

b. Draw stage 1 and stage 3 over the respective labels above.

c. What stays the same from stage to stage in the pattern?

d. Suppose you are looking at stage 9 in the pattern. How many tiles does this stage have?

e. Why do you think your answer in (d) is correct?

Name_____

3. The illustration below shows stage 2 in a growing pattern of square tiles in the shape of the capital letter X. Suppose that each stage has 4 more tiles than the preceding stage.

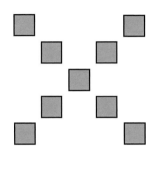

Stage 1   Stage 2   Stage 3

a. Draw stage 1 above the label.

b. Draw stage 3 above the label.

c. What changes from one stage to the next?

What stays the same?

d. Suppose you are looking at stage 8 in the pattern. How many tiles does this stage have?

e. Why do you think your answer in (d) is correct?

Name_____

4. The illustration below shows stage 3 in a rectangular pattern of black and white square tiles. Suppose that all stages in the pattern have a height of 3 tiles and a black tile "core."

Stage 1   Stage 2   Stage 3

a. Draw stage 1 and stage 2 above the respective labels.

b. What characteristics change from one stage to the next in the pattern?

What characteristics stay the same?

c. Suppose you are looking at stage 11 in the pattern. How many tiles (black and white tiles together) does this stage have in all?

How many black tiles does this stage have?

How many white tiles does it have?

d. Why do you think your answers in (c) are correct?

Take a Look at Those Squares

Name_____

The triangle on the right is a *right triangle*. One of its angles is a right angle, which measures 90 degrees. The side that is opposite the right angle is called the *hypotenuse*. Each of the other two angles in the triangle is less than 90 degrees, and the sides that are opposite them are called the *legs* of the triangle.

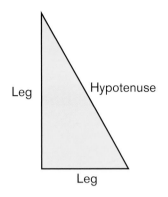

You could use the hypotenuse to form one side of a square adjacent to the triangle, and you could construct squares on the legs of the triangle in the same way, as in the diagram in step 1 below.

   1. Consider the right triangle pictured and the squares constructed on each of its three sides.

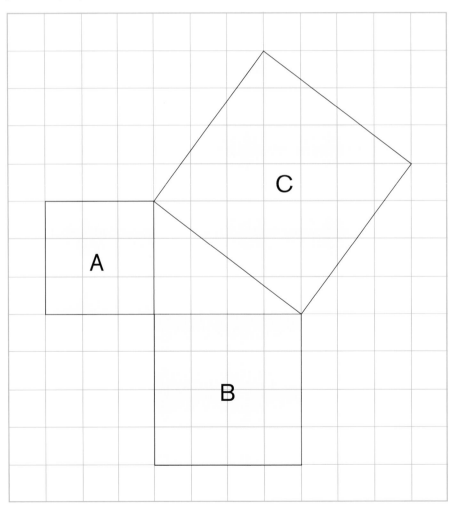

   a. What is the area of each square? (*Hint:* If you can't find the area of a square directly and immediate by counting grid units, then partition (or *decompose*) the square into triangles or other shapes whos

Name_____

areas you can find easily by using grid lines and grid units. Remember that the formula for the area of a triangle is $A = \frac{1}{2}\,bh$.

Area of A = _____ units^2

Area of B = _____ units^2

Area of C = _____ units^2

b. What is the side length of each square?

Side length of A = _____ units

Side length of B = _____ units

Side length of C = _____ units

c. What do you notice about the areas of the three squares?

2. Consider a different right triangle and the squares constructed on its sides.

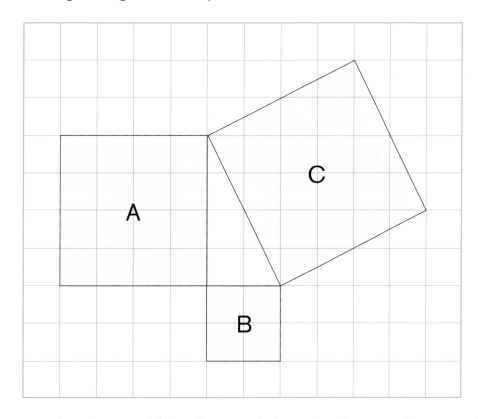

a. What is the area of each square? (*Hint:* If you can't determine the area of a square directly and immediately by counting grid units, decompose the square in the way that you did in step 1.)

Name_____

Area of A = _____ units^2

Area of B = _____ units^2

Area of C = _____ units^2

b. What is the side length of each square? (If the side length is not a whole number of linear grid units, give the measurement to the nearest tenth of a unit.)

Side length of A = _____ units

Side length of B = _____ units

Side length of C ≈ _____ units

c. What do you notice about the areas of the three squares?

3. Consider the right triangle below and the squares constructed on its sides.

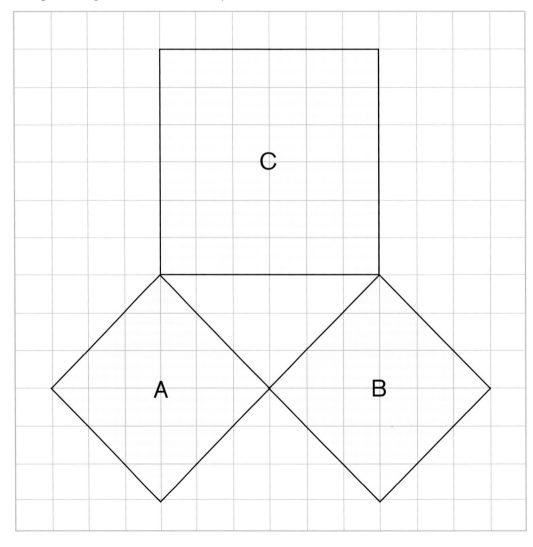

Name_____

a. What is the area of each square? (If you can't count grid units easily, use the same strategy as before.)

Area of A = _____ units^2

Area of B = _____ units^2

Area of C = _____ units^2

b. What is the side length of each square? (If the side length is not an even number of grid units, give the measurement to the nearest tenth of a unit.)

Side of A = _____ units

Side of B ≈ _____ units

Side of C ≈ _____ units

c. What do you notice about the areas of the three squares?

4. Consider the right triangle on the following page and the squares constructed on its sides.

a. What is the area of each square? (Use the same strategy as before.)

Area of A = _____ units^2

Area of B = _____ units^2

Area of C = _____ units^2

b. What is the side length of each square (to the nearest tenth of a grid unit)?

Side length of A ≈ _____ units

Side length of B ≈ _____ units

Side length of C ≈ _____ units

c. What do you notice about the areas of the three squares?

Name_____

5. A screen image from the applet Squaring the Triangle appears at the top of the next page. The right-hand side of the screen shows a right triangle *ABC* with a square on each of its three sides. The left-hand side of the screen presents data on the triangle. Leg *AC* of the pictured triangle measures 9 linear units on the grid, leg *BC* also measures 9 units, and *AB*, the hypotenuse, measures 12.728 units. The displayed data also include measurements for the angles of the right triangle and the areas of the squares on its sides. All values of length represent linear units, and all values of area represent square units.

Take a Look at Those Squares (continued)

Name_____

ote the two buttons that appear elow the image of triangle and ie squares. By sliding these buttons, applet users can adjust the ·ngths of legs *BC* and *AC.* At the ame time, they will change the ngth of the hypotenuse *AB,* the ·ea of right triangle *ABC,* and the ·eas of the squares on its sides. he screen will show the new right iangle and squares, as well as ie accompanying data.

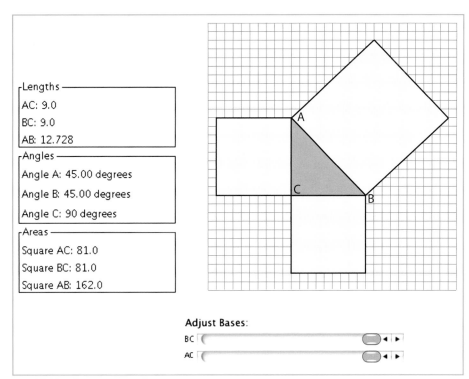

Lengths
AC: 9.0
BC: 9.0
AB: 12.728

Angles
Angle A: 45.00 degrees
Angle B: 45.00 degrees
Angle C: 90 degrees

Areas
Square AC: 81.0
Square BC: 81.0
Square AB: 162.0

Adjust Bases:
BC
AC

a. Use the applet to explore different right triangles and their corresponding squares. Keep records on five of your triangles in the table below. *If you don't have access to the applet,* draw five different right triangles on grid paper. Position all vertices at intersections of grid lines, and label the vertices *A, B,* and *C,* with *AB* as the hypotenuse, as in the screen image from the applet. Complete the chart below for the triangles that you have drawn.

Lengths (linear grid units)			Areas (square grid units)		
AC	*BC*	*AB*	$(AC)^2$	$(BC)^2$	$(AB)^2$

Name_____

b. What do you notice about the areas of the three squares in each of your examples?

6. On the basis of all your work up to this point, can you make a general statement about the relationship among the areas of the squares constructed on the two legs of a right triangle and the area of the square constructed on the hypotenuse? Use the diagram and its labels to help you write a clear statement

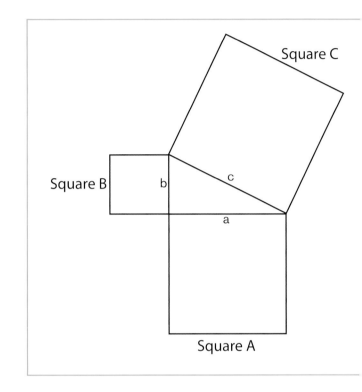

Sets of Right Triangles for "Can You Prove It?"

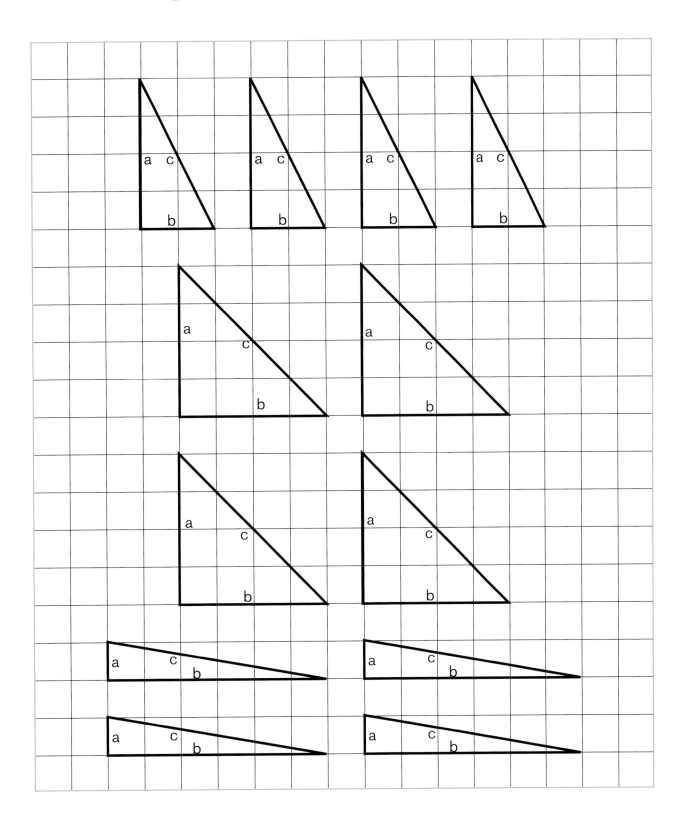

Can You Prove It?

Name_____

1. Cut out each of the three sets of four right triangles on the accompanying sheet, "Sets of Right Triangles for 'Can You Prove It?'"

2. Work with the sets one at a time on a sheet of centimeter grid paper. Arrange each set of four right triangles in such a way that their legs line up with grid lines to form a square whose side length is equal to $a + b$. The hypotenuses of these triangles will "frame" a smaller square with side length is c, as shown at the right.

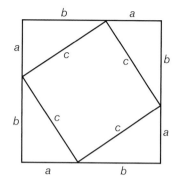

3. Draw each arrangement of triangles on the grid paper, producing three figures that look like the one above but match the triangles that you use to make them. Note that each figure has two squares—an outer, *circumscribed*, square with a side length of $a + b$ and an inner, *inscribed*, square with a side length of c. (Observe that a, b, and c are different lengths in each figure.)

4. Look at the three outer squares (side lengths $a + b$) that you have constructed from the different sets of right triangles. Identify the square formed from the right triangles in which leg a is longer than leg b.

   a. What expression can you write for the area of this outer square if you think about it just as a square with a side length of $a + b$ and ignore everything else in the figure?

   $A =$ _____

   b. You can think about the area of the outer square in another way. What expression can you write for the area of the square if you think about it as the sum of the areas of the four right triangles with legs a and b and the area of the inner square with side length c?

   $A =$ _____

   c. The expressions that you have written in parts (*a*) and (*b*) both give the area of the outer square, so you know that they are equivalent. Set your two expressions equal to each other and simplify them. What remains?

   _____ = _____

Navigating through Problem Solving and Reasoning in Grades 6–

Can You Prove It?

Name_____

5. Examine the outer square in which *a* is equal to *b* in the side length *a* + *b*. Repeat the process from step 4:

   a. What expression can you write for the area of this outer square if you think about it just as a square with a side length of *a* + *b* and ignore everything else in the figure?

   $A =$ _____

   b. What expression can you write for the area of the outer square if you think about it as the sum of the areas of the four right triangles with legs *a* and *b* and the area of the inner square with side length *c*?

   $A =$ _____

   c. Set your expressions from parts (*a*) and (*b*) for the area of the outer square equal to each other and simplify them. What remains?

   _____ = _____

6. Examine the outer square in which *a* is less than *b* in the side length *a* + *b*. Repeat the process from steps 4 and 5:

   a. What expression can you write for the area of this outer square if you think about it just as a square with a side length of *a* + *b* and ignore everything else in the figure?

   $A =$ _____

   b. What expression can you write for the area of the outer square if you think about it as the sum of the areas of the four right triangles with legs *a* and *b* and the area of the inner square with side length *c*?

   $A =$ _____

   c. Set your expressions from parts (*a*) and (*b*) for the area of the outer square equal to each other and simplify them. What remains?

   _____ = _____

7. Compare the results of your work up to this point with the statement that you wrote in step 8 in "Take a Look at Those Squares." How can you relate what you wrote earlier to your discoveries here? Explain your thinking.

Solutions for the Blackline Masters

Solutions for "Sum-thing about Consecutive Numbers

The students identify the counting numbers from 1 to 35 that are sums of consecutive counting numbers. The following chart summarizes the results that they obtain in steps 1 and 2:

Number of Consecutive Addends	2	3	4	5	6	7
Counting Number	3, 5, 7, 9, 11, 13, 15, 17, 19, 21, 23, 25, 27, 31, 33, 35	6, 9, 12, 15, 18, 21, 24, 27, 30, 33	10, 14, 18, 22, 26, 30, 34	15, 20, 25, 30, 35	21, 27, 33	28, 35

Patterns that the students observe may include the following:

- Counting numbers with two consecutive addends are all the odd numbers that are greater than or equal to 3. Adding any two consecutive counting numbers always yields an odd number: $n + (n + 1) = 2n + 1$. Two times any number is even, and adding 1 gives an odd number.

- Counting numbers with three consecutive addends are all the multiples of 3 that are equal to or greater than 6: $n + (n + 1) + (n + 2) = 3n + 3$. This expression describes multiples of 3 that greater than or equal to 6 when n is any counting number.

- Counting numbers with four consecutive addends follow the pattern of $4n + 6$, which is the sum of $n + (n + 1) + (n + 2) + (n + 3)$. The set of these numbers begins with 10 and includes multiples of 4 added to 10.

- Counting numbers with five consecutive addends are multiples of 5 that are equal to or greater than 15: $n + (n + 1) + (n + 2) + (n + 3) + (n + 4) = 5n + 10$.

- Counting numbers with six consecutive addends follow the pattern of $6n + 15$, which is the sum of $n + (n + 1) + (n + 2) + (n + 3) + (n + 4) + (n + 5)$. The set of numbers begins with 21 and includes multiples of 6 added to 21.

- Counting numbers with seven consecutive addends are multiples of 7 that are equal to or greater than 28: $n + (n + 1) + (n + 2) + (n + 3) + (n + 4) + (n + 5) + (n + 6) = 7n + 21$.

It is interesting to note that the set composed of the first number of each sequence ($\{3, 6, 10, 15, 21, 28\}$) is a subset of the triangular numbers. In particular, the set of numbers composed of the sums of two consecutive numbers begins with the second triangular number; the set of numbers composed of the sums of three consecutive numbers begins with the third triangular number, and so on. The chart below shows the sequence of these numbers.

	1st	2nd	3rd	4th	5th	6th	7th	...	nth
Triangular Number	1	3	6	10	15	21	28	...	$\frac{n(n+1)}{2}$

Finding the first number in the sequence of counting numbers that are sums of eight consecutive numbers is a simple matter of using the triangular number "rule" to find the eighth triangular number:

$$\frac{8(8+1)}{2}, \text{ or } 36.$$

The sequence then continues by adding multiples of 8 to 36:

$$\{36, 44, 52, 60, \ldots\}.$$

It is also worth remarking that the only counting numbers that cannot be represented as sums of consecutive numbers are the powers of 2 ({1, 2, 4, 8, 16, 32,...}). Investigating this phenomenon can be a challenging extension of this investigation. For additional observations, see Driscoll (1999).

In steps 3 and 4, the students predict whether each number in a group of five is the sum of two, three, four, or more consecutive counting numbers and discuss any shortcuts that they can suggest for determining which numbers are the sums of two or more consecutive addends. Students' suggestions will vary; they should base their predictions and suggestions on observations like those above. Limiting the investigation of the numbers 45, 57, 62, 75, and 80 to two through seven consecutive addends yields the following results:

45 is the sum of two, three, five, and six consecutive addends:

$22 + 23 = 45$, $14 + 15 + 16 = 45$, $7 + 8 + 9 + 10 + 11 = 45$, $5 + 6 + 7 + 8 + 9 + 10 = 45$

57 is the sum of two, three, and six consecutive addends:

$28 + 29 = 57$, $18 + 19 + 20 = 57$, $7 + 8 + 9 + 10 + 11 + 12 = 57$

62 is the sum four consecutive addends:

$14 + 15 + 16 + 17 = 62$

75 is the sum of two, three, five, and six consecutive addends:

$37 + 38 = 75$, $24 + 25 + 26 = 75$, $13 + 14 + 15 + 16 + 17 = 75$, $10 + 11 + 12 + 13 + 14 + 15 = 75$

80 is the sum of five consecutive addends:

$14 + 15 + 16 + 17 + 18 = 80$

See also the discussion on page 16 of the relationship between the factors of an odd number and the number of consecutive numbers that may sum to that odd number.

Solutions for "Looking for the Least"

1. *a−c.* To have equal numbers of hot dogs and rolls, the smallest number of packages is 4 packages of hot dogs (with 10 in a package) and 5 packages of rolls (with 8 in a package). If everyone wants to eat 2 hot dogs, these packages will feed 20 people. Your students might work with 40 in this way, as the LCM, but other approaches are possible, including thinking of 8 × 10, or 80, dogs-and-rolls, which would feed 40 people when each person gets 2 dogs-and-rolls.

   d−e. Feeding a total 30 people will require 60 dogs-and-rolls, since each person wants 2 hot dogs. The lunch planners must purchase 6 packages of hot dogs and 8 packages of rolls, and they will have 4 rolls left over.

2. *a−b.* The LCM of 15 and 20 is 60; thus, in 60 minutes, the lights will be synchronized again. Because they began their revolutions at 9:00 a.m., they will again be in sync in an hour, at 10:00 a.m. Students may also approach the problem by multiplying: 15 × 20 = 300. The lights will be synchronized in 300 minutes, or 5 hours. This is true, but the lights will also have been synchronized at the end of each hour before 5 hours.

3. *a−b.* The LCM of 9 and 10 is 90 because 9 and 10 are *relatively prime*—that is, they have no common factors except 1. This means that the smallest number of hours that the child and the teenager must sleep for both to have slept the same number of hours is 90 hours. Thus, the child must sleep 9 nights, and the teenager must sleep 10 nights.

   c−d. The LCM of 9, 10, and 8 is 360. This means that the smallest number of hours that the child, the teenager, and the adult must sleep for all three to have slept the same number of hours is 360 hours. Thus, the child must sleep 36 nights, the teenager must sleep 40 nights, and the adult must sleep 45 nights.

Solutions for "In-Venn-stigating Factors and Multiples"

1. *a–b.* The numbers 8 and 3 are *relatively prime* numbers—that is, they share no common factors except 1. This means that common multiples (in the intersection) are the multiples of their product (8 × 3 = 24).

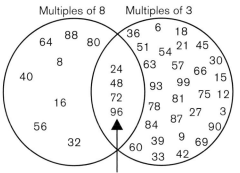

Multiples of 8   Multiples of 3

Multiples of the product of 8 × 3

2. *a–c.* The numbers 2 and 7 are *prime* numbers—thus, they share no common factors except 1. This means that common multiples (in the intersection) are the multiples of their product (2 × 7 = 14). Other common multiples that are less than 100 are 42, 56, 70, 84, and 98.

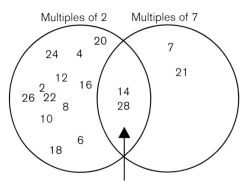

Multiples of 2   Multiples of 7

Multiples of the product of 2 × 7

3. The numbers in the intersection are multiples of the product of the two numbers because the two numbers are either relatively prime or prime and thus have no factors in common except 1. If two numbers *A* and *B* are prime or relatively prime, then the common multiples of *A* and *B* are the product *A* × *B* and its multiples.

4. *a–b.* All the multiples of 15 are multiples of 5 since 5 is a factor of 15; thus, all the multiples of 15 are in the intersection.

   Because 5 is equal to 15 × $\frac{1}{3}$, one third of the multiples of 5 are multiples of 15.

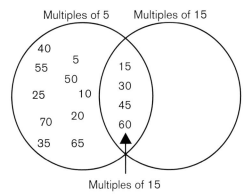

Multiples of 5   Multiples of 15

Multiples of 15

5. *a–c.* All the multiples of 8 are multiples of 4 since 4 is a factor of 8; thus, all the multiples of 8 are in the intersection.

   Because 4 is equal to 8 × $\frac{1}{2}$, one half of the multiples of 4 are multiples of 8. Other common multiples that are less than 100 are 56, 64, 72, 80, 88, and 96.

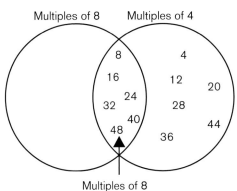

Multiples of 8   Multiples of 4

Multiples of 8

6. *a–d.* In steps 4 and 5, in each pair of numbers, one number is a factor of the other. Given two numbers *A* and *B* with *A* a factor of *B*, $B = n \times A$ for some natural number *n*. As a result, the common multiples of *A* and *B* are all the multiples of *B*. Furthermore, all multiples of *B*, the "nonfactor" number, are multiples of *A*, the given factor, and $\frac{A}{B}$ of the factor's multiples are multiples of the "nonfactor." All multiples of the "nonfactor" number are thus in the intersection of a Venn diagram. The numbers 4 and 24 offer another example of a pair of numbers where one number is a factor of the other: 4 is a factor of 24; $24 = 6 \times 4$. The common multiples of 4 and 24 are all multiples of 24, and $\frac{1}{6}$ of the multiples of 4 are multiples of 24. Pairs of this type and their diagrams contrast with the pairs of the type explored in steps 1 and 2. Those pairs consist of numbers that are prime or relatively prime, with no factor in common except 1. In those cases, the common multiples are multiples of the product of the paired numbers, and all the regions of the Venn diagrams are populated. The numbers 13 and 7 offer another example of a pair of numbers like those in steps 1 and 2; 13 and 7 are both prime numbers.

7. *a–c.* The common multiples of 12 and 8 appear in the intersection. The LCM is the smallest multiple in the intersection; the numbers in the intersection are multiples of the LCM.

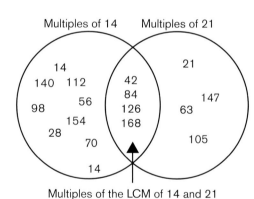

8. *a–c.* The common multiples of 14 and 21 appear in the intersection. The LCM is the smallest multiple in the intersection; the numbers in the intersection are multiples of the LCM.

9. *a–c.* The numbers in the intersection are multiples of the LCM of the two numbers. For a pair of numbers *A* and *B*, if neither is a factor of the other, but *A* and *B* have a greatest common factor that is greater than 1, then the LCM of *A* and *B* is equal to the product of *A* and *B* divided by the greatest common factor. (For example, in the case of 14 and 21 in step 8, the greatest common factor is 7: $14 \times 21 = 294$, and $294 \div 7 = 42$; the LCM is 42.) The common multiples of *A* and *B* are multiples of the least common multiple of *A* and *B*. Two other examples that would produce similar Venn diagrams are 15 and 12 (LCM $= 15 \times 12 \div 3 = 60$) and 24 and 36 (LCM $= 24 \times 36 \div 12 = 72$).

Navigating through Problem Solving and Reasoning in Grades 6–

10. *a–b.* The GCF of 12 and 8 is 4; the Venn diagram appears to the right; note that 4 is the greatest factor in the intersection.

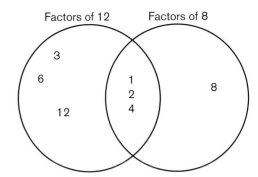

Factors of 12   Factors of 8

11. *a–b.* The GCF of 14 and 21 is 7; the Venn diagram to the right; note that 7 is the greatest factor in the intersection.

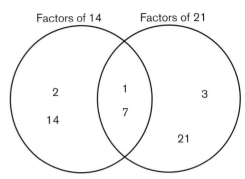

Factors of 14   Factors of 21

12. *a–b.* The LCM of 12 and 8 is 24, which is the product of the prime factors in all the regions of the Venn diagram: $2 \times 2 \times 2 \times 3$. The GCF of 12 and 8 is 4, which is the product of the prime factors in the intersection: 2×2.

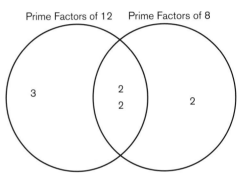

Prime Factors of 12   Prime Factors of 8

13. *a–b.* The LCM of 14 and 21 is 42, which is the product of the prime factors in all the regions of the Venn diagram: $2 \times 3 \times 7$. The GCF of 14 and 21 is 7, which is the prime factor in the intersection.

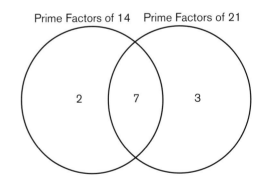

Prime Factors of 14   Prime Factors of 21

Solutions for "What Are the Relationships?"

1–4. Students should conjecture that the product of each number pair is equal to the product of its GCF and LCM. They can argue that identifying the prime factorizations facilitates a determination of both the GCF and the LCM for the number pair. They can use a Venn diagram to organize the prime factors of each pair of numbers. The product of the prime factors in the intersection of the Venn diagram represents the GCF of the numbers. The product of all the prime factors in the union of the sets shown in the Venn diagram represents the LCM for the number pair. The product of the GCF and the LCM is the same as the product of the number pair because all the factors in the product of the numbers themselves appear in the product of the GCF and the LCM.

Solutions for "Parallel(ogram) Universe"

1. *a–b.* Students' methods of determining the areas of the parallelograms will vary; figures 2.1*a* and 2.5 illustrate two possible methods. The areas of the parallelograms are (*a*) 35 square units and (*b*) 27 square units.

2. Students' descriptions of their methods for determining the areas will vary. For example, students who used the method shown in figure 2.5 might say, "I made a rectangle by cutting a small triangle from each side of the parallelogram and rotating it to form a rectangle."

3. Students will give various explanations of their processes to a hypothetical fifth grader who knows how to find the area of a rectangle. A sample follows: "Make a rectangle out of the parallelogram by cutting and moving small triangles on each side of it. The rectangle will have the same area as the parallelogram. Then use the formula for the area of rectangle to find the area of the parallelogram."

4. Students' formulas may vary in expression, but all should communicate the same idea: Area = base × height.

Solutions for "Moving into Triangular Territory"

1. *a–c.* Students' methods of determining the areas of the triangles will vary. Figure 2.7 illustrates two methods that they might use for the isosceles right triangle in (*a*); they can apply the second method shown in the figure to any triangle, including the acute triangle in (*b*) and the obtuse triangle in (*c*), both of which are scalene.

2. Students' descriptions of their methods will vary.

3. Students will give various explanations of their processes to anyone who knows how to find the areas of a rectangle and a parallelogram. A sample follows: "Make the triangle half of a parallelogram that uses two sides of the triangle as two of its sides. Then figure out the area of the parallelogram and divide it by 2."

4. Students' formulas may vary in expression, but all should communicate the same idea:
Area = $\frac{1}{2}$(base × height).

Solutions for "How Much Area Does a Trapezoid Trap?"

1. *a–b.* Students' methods of determining the areas of the trapezoids will vary. Figure 2.9 illustrates two methods that they might use for the isosceles trapezoid in (*a*). Figure 2.10 illustrates two methods that they can apply to the non-isosceles trapezoid in (*b*), and figure 2.12 illustrates five other methods that they can use with any trapezoid.

2. Students' answers will vary, depending on the method that they used.

3. *a–c.* Students' methods of determining the areas will vary; the area of (*a*) is $\frac{1}{2}(14+6)(5)$, or 50, square units, the area of (*b*) is $\frac{1}{2}(11+3)(8)$, or 56, square units, and the area of (*c*) is $\frac{1}{2}(14+2)(6)$, or 48, square units.

4. See fig. 2.12 and the accompanying discussion in the text for possible expressions of formulas depending on the methods that the students use. Although the expressions may vary, all should communicate the same idea:
Area = $\frac{1}{2}$ (base 1 + base 2) × height.

Solutions for "Carryover to Kites"

1. *a–b.* Students' methods of determining the areas of the two kites will vary. Figure 2.13 shows four methods. The areas of the kites shown are (*a*) $\frac{1}{2}((6 \times 3) + (6 \times 4))$, or 21, square units and

   (*b*) $2\left(\frac{1}{2}(8 \times 3)\right)$, or 24, square units.

2. Students' explanations will vary, but all students should have found methods that they could use to find the area of any kite.

3. Students' methods of determining the areas of the given kites will vary. The areas follow:

   a. 32 square units

   b. 26 square units

   c. 70 square units

4. Students' expressions of the formula may vary, but all should communicate the same idea:

   Area $= \frac{1}{2}$ (diagonal 1 $\times$ diagonal 2).

Solutions for "Stepping Up to Stairs"

1. Students' methods of determining the areas of the two "stairs" will vary. Figure 2.14 shows three ways to manipulate stair 1 to determine its area efficiently. The areas of the stairs shown are (*a*) $\frac{1}{2}(8 \times (8+1))$, or 36, square units and (*b*) $\frac{1}{2}(7 \times (3+15))$, or 63, square units.

2. Students' expressions of the formula may vary; one possible expression is discussed on pages 49–51.

Solutions for "Diagonal Discoveries"

1. The students should cut out the five rectangles with special care.

2. The completed table appears below (it also appears as fig. 3.2):

Rectangle	Length (cm)	Width (cm)	Perimeter (cm)	Area (cm²)	$\dfrac{\text{Width}}{\text{Length}}$
Rectangle 1	5	4	18	20	$\frac{4}{5}$
Rectangle 2	8	6	28	48	$\frac{6}{8}$, or $\frac{3}{4}$
Rectangle 3	15	12	54	180	$\frac{12}{15}$, or $\frac{4}{5}$
Rectangle 4	18	14	64	252	$\frac{14}{18}$, or $\frac{7}{9}$
Rectangle 5	20	16	72	320	$\frac{16}{20}$, or $\frac{4}{5}$

3. Students' responses will vary. Rectangles 1, 3, and 5 are similar rectangles. The side lengths of rectangle 3 are proportional to the corresponding side lengths of rectangle 1—specifically, they are three times those of rectangle 1. The side lengths of rectangle 5 are also proportional to the corresponding side lengths of rectangle 1—specifically, they are four times those of rectangle 1. The angles of rectangles 1–5 are obviously congruent; the students know that every angle in a rectangle is 90 degrees. However, they should be aware that corresponding angles are congruent in all similar figures.

4. If necessary, assist the students in the process of drawing the diagonal from the bottom left angle vertex (*A*) to the top right vertex (*C*) on each rectangle and aligning all vertices *A*, sides *AB*, and sides *AD* so that the five rectangles are "nested."

5. Students' responses will vary, depending on the relationships that they find among the diagonals of the rectangles. Ideally, they should notice that the diagonals for the similar rectangles (1, 3, and 5) are collinear, but those for the other two rectangles (2 and 4) are not. Representing the rectangles on graph paper may promote students' conclusions that the diagonals of similar rectangles that are "nested" lie in the same line.

6. Students' generalizations about the relationship among the diagonals and their formulations of the generalizations will vary. A sample conjecture follows: "If two or more 'nested' rectangles are similar, then their diagonals are collinear."

7. Students' mathematical arguments will vary. Some might work on grids and reason that the diagonals have the same slope and *y*-intercept, so the diagonals are indeed collinear. They might also notice that the slope of the line is the same as the simplified ratios for the sides of the rectangles. Other students might apply an underlying principle of dilations—that a dilation extends all points in a figure a certain distance (determined by scale factor) through a particular point. Aligning the rectangles with a common vertex provides a center of dilation not only for the sides of the rectangles but also for the diagonals.

Solutions for "Who Is Most Popular"

1. Students' choices of the three most popular celebrities will vary, depending on the criteria for their selections. See the discussion on pages 80–81.

2. Students' explanations of their choices will vary but should take account of the context and any assumptions that they make.

Solutions for "How Do the Dollars Stack Up?"

1. Students' descriptions of the variability of the celebrities' earnings will differ but should take note of the clustering of the data in the region between approximately $5 million and $65 million, with a "peak" around $30 million—about $10 million less than the median of the data set ($41 million) and about $35 million less than the mean ($68.2 million). They should note that twelve celebrities, or 60 percent of them, earned less than $50 million; five celebrities, or 25 percent, earned between $50 million and $100 million; and only three celebrities, or 15 percent, earned $100 million or more.

2. Students' responses will vary but should note that if the outliers were filtered out, the mean and the median would more closely approximate the "peak" that the students see in the clustered data. Removing $185, $225, and $290 million (the earnings of Mel Gibson, Oprah Winfrey, and George Lucas, respectively) from consideration reduces the mean from $68.2 million to $39.1 million and the median from $41 million to $34 million. Regarding $80 and $87 as "on the tail" also and filtering them out as well reduces the mean and median still more, making the mean $33.2 million and the median $33 million.

Solutions for "Word Spreads"

1. Students' responses will vary but should note that eight celebrities out of the twenty, or 67 percent, had no cover stories in magazines in the period of the data. Zero is by far the most common value. The values tend to cluster

Navigating through Problem Solving and Reasoning in Grades 6–

between 0 and 6, and 9 (the number of cover stories garnered by Julia Roberts) appears to be an outlier. Be sure that students understand that decimal fractions in data on magazine cover stories reflect stories that celebrities "shared" with others.

2. Students' responses will vary but should note that the numbers of newspaper articles that mention celebrities' names tend to cluster between approximately 0 and 22,000, with fourteen celebrities, or 70 percent, mentioned in fewer than 15,000 newspaper articles, and nine celebrities, or 45 percent, mentioned in fewer than 10,000 articles. The students might regard the two values that appear at the "top" of the graph (80,075 articles that mention Bill Clinton and 48,467 articles that mention Tiger Woods) as outliers. Be sure that students think about the fact that the data count newspaper articles in which a celebrity's name appears even just one time.

3. Students' responses may vary, but they should regard the two data points with the greatest "y-values" (number of newspaper articles) and the single data point with the greatest "x-value" (number of magazine cover stories) as the outliers in this graph. They should have noted that they are examining x-values when they consider the variability in the numbers of cover stories about celebrities in magazines and y-values when they consider the variability in the numbers of newspaper stories that mention their names. Filtering out the outliers can make the clustering and spread of the remaining data easier to analyze, but if the students are attempting to identify the celebrities with the most print publicity, they would not want to filter out these data points but instead inspect them closely.

4. Students' selections of the celebrity with the most publicity in print and their explanations of their choices may vary, but many are likely to grant Bill Clinton this distinction. With approximately 80,000 newspaper articles in which his name appeared, he is far ahead of the other celebrities in this category. On the basis of this datum, he is likely to get the nod, despite the fact that his number of magazine cover stories is smaller than the numbers for seven of the celebrities.

Solutions for "Change the Data–Change the Stats?"

1. *a*. The students suppose that the two students who originally recorded 11 hours of TV watching actually watched 13 hours apiece during the week. Thus, in the graph of the data, your students will remove the two "blocks" that make up the bar above 11 and add them to the four blocks that make the bar above 13. They should predict that this change will increase the mean slightly but keep the median unchanged (since the change is entirely on one side of the median).

   b. When the students make the change, they should see that the mean changes from 7.4419 to 7.5349 while the median stays the same.

2. *a*. The students now suppose that three of the five students who originally recorded 8 hours of TV watching actually watched 5 hours apiece during the week. Thus, working with the new graph from step 1, your students will remove three blocks from the bar above 8 and add them to the bar above 5. The students should predict that this change will decrease the mean—and the median as well, since this shuffling and changing of data shifts the middle value of the set—the value that divides the set in half.

   b. If the students remove and add blocks one at a time, they will see the incremental change in the mean: 7.5349 to 7.4651 to 7.3953 to 7.3256. When they remove the second block from the bar above 8 and add it to the bar above 5, they will see the median change from 7 hours to 6 hours.

3. Students' explanations of why the median changed in step 2 will vary. The data after step 1 and before step 2 were as follows:

   0 0 0 1 3 3 4 4 5 5 5 5 5 6 6 6 6 6 6 6 7 **7** 7 7 7 7 8 8 8 8 8 9 13 13 13 13 13 13 14 14 15 15

   Note that the twenty-second value—the middle value in the string of forty-three values—is a 7. It is the second 7 in the string, highlighted in green. Three times, the students remove an 8 and add a 5. After they remove the first 8 and add the first 5, the data string changes as shown. The highlighting continues to show the middle value, which is now the first 7 in the string.

   0 0 0 1 3 3 4 4 5 5 5 5 5 6 6 6 6 6 6 6 **7** 7 7 7 7 7 8 8 8 8 9 13 13 13 13 13 13 14 14 14 15 15

After the students remove the second 8 and add the second 5, the data string changes again, as shown. The highlighting continues to show the middle value, which is now the last 6 in the string.

0 0 0 1 3 3 4 4 5 5 5 5 5 5 6 6 6 6 6 6 **6** 7 7 7 7 7 7 8 8 8 9 13 13 13 13 13 13 14 14 14 15 15

After the students remove the third 8 and add the third 5, the data string changes as shown, with the highlightir showing the middle value, which is still a 6 but now the second-to-last 6 in the string. Because of the presence of *repeated values*, the median does not appear to change this time.

0 0 0 1 3 3 4 4 5 5 5 5 5 5 5 6 6 6 6 6 6 **6** 6 7 7 7 7 7 7 8 8 9 13 13 13 13 13 13 14 14 14 15 15

4. To help your students visualize the problem, you might present 12 boxes:

Explain that to solve the problem, the students can fill in the 12 boxes with whole numbers. The values can increase from left to right, or right to left. If they increase from left to right, box 12 must contain an 8, and the values in boxes 6 and 7 must total to 10. The sum of all 12 values must be 60. Such a dissection of the problem can be helpful to middle school students.

a. Students can use different approaches in changing five values in the data set to make the mean greater than 5 hours while letting the median remain 5 hours and the maximum remain 8 hours. The students should understand that the median divides a data set in such a way that half the values are at or below the median and half are at or above it, and the median itself may or may not be a value in the data set. In the students' case, the set has an even number of data values, so the median is the average of the middle two data values. Because the median must continue to be 5 hours, these two values must continue to add to 10, for an average 5. One way to change the mean but not the median would be to increase each of the top five data values by 1 hour. This process would leave the sixth and seventh data values unchanged, along with the median. However, because the sum of data values increased, the mean would also increase and thus would be greate than 5 hours. Alternatively, the students could change five of the values in the set by adding to some of them and subtracting from others, while being sure that the total amount added was greater than the total amount subtracted and that the adjusted list still fulfilled the conditions that the maximum value was 8 and the sixth and seventh values totaled to 10. You might pose the following challenge question to your students: "How would the task change if you were interested only in *changing* the mean in either direction—decreasing it or increasing it?"

b. Students' new sets and explanations will vary. When the students make one of the two smallest data values greater than the median in their new 12-value set from part (*a*) (with a median of 5), they will increase the mean, and they may or may not increase the median, depending on what happens to the middle values—the sixth and seventh values—as a result of their change. For example, suppose that the set that a student originally composes in step 4 consists of

2 2 2 4 4 4 6 6 6 8 8 8,

for a mean of 5, a median of 5, and a maximum value of 8. Also suppose that the student's new set in part (*a* of step 4 consists of the following values:

2 2 2 4 4 4 6 7 7 9 9 9.

The mean is now approximately 5.42—greater than 5, as part (*a*) of step 4 stipulates—while the median continues to be 5. The sixth and seventh values are a 4 and a 6, which add to 10 and average to 5. In this case, taking one of the two smallest data values in the set—one of the 2s—and boosting it above the median means making it greater than or equal to 6—for example,

2 2 4 4 4 6 6 7 7 9 9 9.

Now the mean increases to 5.75, and the median increases to 6. However, consider a different example. Suppose that another student originally composes the following set in step 4:

2 2 2 3 5 5 5 5 7 8 8 8.

This set has a mean of 5, a median of 5, and a maximum value of 8, as required. Suppose that this student's new data set in part (*a*) of step 4 consists of the following values:

3 3 3 4 5 5 5 5 8 8 8 8.

The mean is now approximately 5.42—greater than 5, as part (*a*) of step 4 stipulates—while the median continues to be 5. In this case, taking one of the two smallest data values in the set—one of the 3s—and boosting it above the median means making it greater than 5—for example,

3 3 4 5 5 5 5 6 8 8 8 8.

In this case, the mean increases, becoming approximately 5.67, but the median remains the same—5.

c. Students' responses and explanations will vary, depending on their sets, but they should recognize that increasing the maximum value in a data set, even by a large amount, does not affect the median.

5. *a–b.* Students' estimates of the location of the median and the relative position of the mean (greater than, the same as, or less than the median) in the set shown in the graph below will vary.

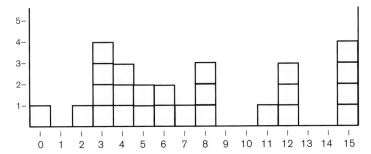

However, a visual inspection of the distribution of the data should help students guess that both measures of center are in the neighborhood of 6 or 7 hours. They may also suspect that the mean is greater than the median by taking into consideration the maximum value of the set and the number of data values that are at or near the maximum. In fact the median of the data set is 6 hours and the mean is 7.44.

Solutions for "Growing in Front of Your Eyes"

1. *a–c.* Students should have little trouble in producing stage 4 (shown with stages 1–3 below).

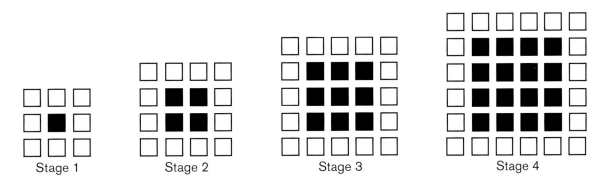

Stage 1 Stage 2 Stage 3 Stage 4

The students will probably see that the side of the large square in the pattern grows by one tile from one stage in the pattern to the next. So they will draw stage 4 with six white tiles on a side. They will draw these white tiles as a "frame" around a square inner "core" of black tiles, with four tiles on a side of this core. Or they may start by inspecting the black tiles, recognizing the pattern in them easily and so drawing them first, and then drawing a frame of white tiles around them. Either way, they should come up with a total of 36 tiles, consisting of 20 white tiles and 16 black tiles. Their explanations of how they knew to draw these numbers of tiles will vary.

2. *a–c.* Stage 6 has 64 tiles in all, consisting of a "frame" of 28 white tiles around a square "core" of 36 black tiles. Students' explanations of how they arrived at these counts will vary. The students should reason from the pictures of stages 1–3 and their drawing of stage 4 through stage 5 and to stage 6. They should be able t use their techniques from step 1, where they drew stage 4, to visualize stage 6 in their minds. They may recognize that the side of the large square (white frame plus inner square core) goes from six tiles at stage 4 seven tiles at stage 5 to eight tiles at stage 6, so stage 6 has 8 × 8, or 64 tiles in all. They may also reaso that the black tiles go from 16 at stage 4 to 25 at stage 5 to 36 at stage 6, so stage 6 has (64 − 36), or 2 white tiles.

3. *a–c.* A sample completed table appears in figure 5.4.

4. Stage 17 has an odd number of black tiles. Students should note that the square "core" of black tiles in any odd-numbered stage is made up of an odd number of tiles. The side of this square core also consists of an odd number of tiles, and the students should observe, if they do not already know, that the square of an odd number is also odd (the product of any two odd numbers is odd).

5. *a–c.* Stage 10 in the pattern has 44 white tiles, and at this stage the pattern has 100 black tiles and 144 tiles altogether. Students can approach the problem of determining the stage that has 44 white tiles in different ways. If they were unsuccessful in developing an algebraic expression for the number of white tiles at stage n, or if they are unable to use the algebraic expression directly in an equation ($4n + 4 = 44$; $n = 10$), they might look at the progression of white tiles in the pattern from stage to stage and make a little table:

Stage	1	2	3	4	5	6	7	8	9	10
White tiles	8	12	16	20	24	28	32	36	40	44

Alternatively, they might use a trial-and-error approach, starting with the number of a stage, then considering how many black tiles would be in the square "core" at this stage, and finally considering how many tiles they would need to "frame" this core. For example, they might guess that stage 8 has 44 white tiles. But they can reason that stage 8 in fact has 64 black tiles, because the number of the stage is equal to the number of black tiles that the square core has on a side, and 8 × 8 = 64. To frame this core, the students would need 8 + 8 + 10 + 10, or 36, white tiles. Next, they might guess that stage 9 has 44 white tiles. Bu in this case they can reason that stage 9 has 81 black tiles, because 9 × 9 = 81. To frame this core, the students would need 9 + 9 + 11 + 11, or 40, white tiles. Finally, they might guess stage 10 has 44 white tiles and reason that stage 10 would have 100 black tiles, and to frame this core, they would need 10 + 10 + 12 + 12, or 44, white squares.

6. *a–c.* The students know from their work in step 2 that stage 6 has a total of 64 tiles, and the table that they completed in step 3 shows them that stage 5 has 25 black tiles, 24 white tiles, and 49 tiles in all. The total number of tiles at every stage is a square number (1, 4, 9, 16,…), and no square number exists between 49 and 64. So it is not possible for any stage in the pattern to have a total of 52 tiles. Likewise, it is not possible for any stage to have 52 black tiles, because the numbers of black tiles are also square numbers. However, it is possible for a stage in the pattern to have 52 white tiles. If students could not develop an algebraic expression for the number of white tiles or cannot use the expression successfully to establish this fact ($4n + 4 = 52$; $n = 12$), they can work with the progression of white tiles as in step 5.

Stage	1	2	3	4	5	6	7	8	9	10	11	12
White tiles	8	12	16	20	24	28	32	36	40	44	48	52

Students might note, or you might help them observe, that the number of white tiles is always a multiple of 4. In fact, any multiple of 4 that is greater than or equal to 8 can be the number of white tiles in a frame around a core of black tiles in the pattern.

Solutions for "Change the Pattern–Change the Growth?"

1. *a–c.* Students should have little trouble in producing stage 4 (shown with stages 1–3 below).

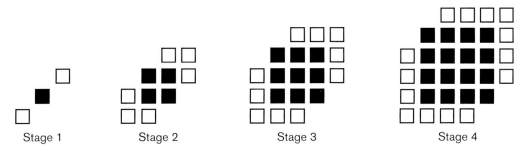

| Stage 1 | Stage 2 | Stage 3 | Stage 4 |

The students will probably draw the inner black-tile "core" first and then draw the outer white tiles, noticing that this time, however, unlike in the first pattern, they must "omit" three tiles at or adjacent to each of two opposite corners of the square core of black tiles. They should come up with a total of 30 tiles at stage 4, consisting of 14 white tiles and 16 black tiles. Their explanations of how they knew to draw these numbers of tiles will vary.

2. *a–c.* Stage 5 has 43 tiles in all, consisting of 18 white tiles and 25 black tiles. Students' explanations of how they arrived at these counts will vary. They should reason from the pictures of stages 1–3 and their drawing of stage 4 to imagine stage 5.

3. *a–c.* A sample completed table appears on the next page.

4. The students should not have a great deal of difficulty in deciding that stage 20 has an even number of black tiles. They can see that the pattern of growth in the black tiles in this pattern is the same as in the previous one. They should note that the square core of black tiles at any even-numbered stage is made up of an even number of tiles. The side of this square core also consists of an even number of tiles, and the students should observe, if they do not already know, that the square of an even number is also even (the product of any two even numbers is even).

5. *a–c.* The stage that has 46 white tiles is stage 12, and this stage has 144 black tiles and 190 tiles altogether. Students can approach the problem of determining the stage in different ways. If they were unsuccessful in developing an algebraic expression for the number of white tiles at stage n, or if they are unable to use the algebraic expression directly in an equation ($4n - 2 = 46$; $n = 12$), they might look at the progression of white tiles from stage to stage in the pattern and make a little table:

Stage	1	2	3	4	5	6	7	8	9	10	11	12
White squares	2	6	10	14	18	22	26	30	34	38	42	46

6. *a–c.* The students know that the total number of tiles at any stage of the pattern is the sum of the number of white tiles and the number of black tiles at that stage. Moreover, they know that the number of black tiles is the square of the number of the stage. So they can use a table like that in the solution to step 5 above, which shows numbers of white tiles at stages 1–12 and consider square numbers of black tiles in a guess-and-check approach. For example, stage 8 has 64 black tiles and 30 white tiles, for a total of 94 tiles. Stage 9 has 81 black tiles and 34 white tiles, for a total of 115 tiles. So it is possible for a stage to have a total of 115 tiles—stage 9 does. However, it is not possible for any stage to have 115 black tiles, because 115 is not a square number. The first square number after 100 (10^2) is 121 (11^2). Furthermore, it is not possible for any stage to have 115 white tiles. The students do not need to have succeeded in developing the algebraic expression $4n - 2$ for the number of white tiles at stage n to demonstrate that no stage with 115 white tiles exists—they know that the number of white tiles at any stage is always an even number: the number of white tiles is even at stage 1 (2 white tiles), and it increases by the even number 4 from one stage to the next.

A sample completed table for step 3 below:

Stage	Number of Black Tiles—Describe What You See	Number of White Tiles—Describe What You See	Total Number of Tiles—Describe What You See
1	A 1 × 1 square made up of **1 black tile**	"Clips" at two opposite corners made up of $(1 + 0) \times 2 =$ **2 white tiles**	A 1 × 1 black square with "clips" consisting of 2 white tiles at opposite corners, for **3 tiles in all**
2	A 2 × 2 square made up of **4 black tiles**	"Clips" at two opposite corners made up of $(2 + 1) \times 2 =$ **6 white tiles**	A 2 × 2 black square with "clips" consisting of 6 white tiles at opposite corners, for **10 tiles in all**
3	A 3 × 3 square made up of **9 black tiles**	"Clips" at two opposite corners made up of $(3 + 2) \times 2 =$ **10 white tiles**	A 3 × 3 black square with "clips" consisting of 10 white tiles at opposite corners, for **19 tiles in all**
4	A 4 × 4 square made up of **16 black tiles**	"Clips" at two opposite corners made up of $(4 + 3) \times 2 =$ **14 white tiles**	A 4 × 4 black square with "clips" consisting of 14 white tiles at opposite corners, for **30 tiles in all**
5	A 5 × 5 square made up of **25 black tiles**	"Clips" at two opposite corners made up of $(5 + 4) \times 2 =$ **18 white tiles**	A 5 × 5 black square with "clips" consisting of 18 white tiles at opposite corners, for **43 tiles in all**
6	A 6 × 6 square made up of **36 black tiles**	"Clips" at two opposite corners made up of $(6 + 5) \times 2 =$ **22 white tiles**	A 6 × 6 black square with "clips" consisting of 22 white tiles at opposite corners, for **58 tiles in all**
⋮	⋮	⋮	⋮
n	An $n \times n$ square made up of n^2 **black tiles**	"Clips" at two opposite corners made up of $(n + (n - 1)) \times 2$ $= (2n - 1) \times 2$ $= (4n - 2)$ **white tiles**	An $n \times n$ black square with "clips" consisting of $(4n - 2)$ white tiles at opposite corners, for $n^2 + 4n - 2$ **tiles in all**

Solutions for "What Changes? What Stays the Same?"

1. The students see stage 2 and draw stages 1 and 3 in a pattern based on the shape of the capital letter H.

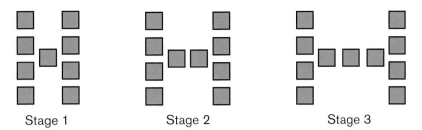

Stage 1	Stage 2	Stage 3

Because stage 1 has 9 tiles and stage 3 has 11 tiles, the students can reason that the only part of the H that is changing from stage to stage is the cross bar, whose number of tiles is equal to the number of the stage.

b. The students should draw stage 1 with 1 tile in the cross bar and stage 3 with 3 tiles in the cross bar.

c. The students should observe that what stays the same is the number of tiles on the two sides. Each side always has 4 tiles.

d−e. The students should observe that the number of tiles in the cross bar is equal to the number of the stage, and the number of tiles on the sides is 2×4, or 8. So the number of tiles at stage 10 is $10 + 8$, or 18. The number of tiles at stage n is $8 + n$.

2. The students see stage 2 and draw stages 1 and 3 in a pattern based on the shape of the letter S.

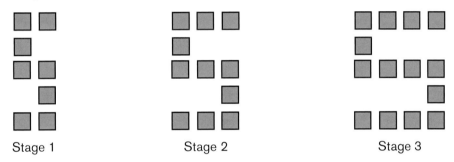

Stage 1	Stage 2	Stage 3

a. The students might think of the Ss in the pattern in terms of rows or columns. They know that stage 1 has 8 tiles and stage 3 has 14 tiles. If they think in terms of rows, they might say that rows 1, 3, and 5 are changing from stage to stage. At stage 1, each of these rows has only 2 tiles, and the S has $6 + 2$, or 8, tiles in all. At stage 3, each of these rows has 4 tiles, and the S has $12 + 2$, or 14, tiles in all. Alternatively, if they think in terms of columns, the students might say that the number of "interior" columns—columns between column 1 and the last column on the right—is the changing feature of the Ss in the pattern. Each interior column always has 3 tiles—one at the top, one in the middle, and one at the bottom. At stage 1, the number of interior columns is 0; at stage 2, it is 1; and at stage 3, it is 2.

b. Students should draw stage 1 with no columns in the interior and stage 3 with two columns, or 6 tiles, in the interior—2 on the bottom, 2 in the middle and 2 on the top.

c. See the solution for part (*a*). The students might say that what stays the same is the number of tiles that separate the "top," "middle," and "bottom" groups of tiles in an S—the tiles in rows 2 and 4. The number of tiles in each of these rows is always 1. Alternatively, they might observe that an S at stage n has $n + 1$ columns, and what stays the same are the numbers of tiles in columns 1 and $n + 1$. The number of tiles in each of these columns is always 4.

d−e. The number of tiles at stage 9 is 32. If students think about the S at stage 9 in terms of rows, they might reason that the number of tiles in each row for rows 1, 3, and 5 is equal to 10, or the number of the stage plus 1, while the number of tiles in each of rows 2 and 4 continues to be 1. So the number of tiles at stage 9 is equal to $3(10) + 2$, or 32. If they think about the S at stage 9 in terms of columns, they might reason that stage 9 has $9 + 1$, or 10 columns, and the number of tiles in each of columns 1 and 10 is 4. The number of "interior columns" is equal to $9 − 1$, or 8. Each of these columns has 3 tiles. Thus, the number of tiles at stage 9 is $8 + 3(9 − 1)$, or 32.

3. The students see stage 2 and draw stages 1 and 3 in a pattern based on the shape of the letter X. They suppose that each stage has 4 more tiles than the preceding stage.

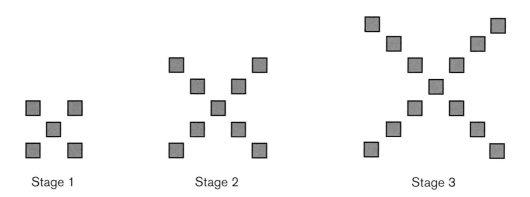

| Stage 1 | Stage 2 | Stage 3 |

a. Each "arm" at stage 1 is 1 tile shorter than each arm at stage 2.

b. Each arm at stage 3 is 1 tile longer than each arm at stage 2.

c. The students might observe that what changes from stage to stage is the number of tiles in the four arms of the X, and what stays the same is the 1 tile at the intersection of all the arms.

d–e. Each stage has one tile in the center in addition to n tiles in each of the four arms. Stage 1 has $1 + 1(4)$, or 5 tiles. Stage 2 has $1 + 2(4)$, or 9 tiles. Stage n has $1 + n(4)$ tiles. So stage 8 has $1 + 8(4)$, or 33 tiles.

4. The students see stage 3 and draw stages 1 and 2 in a sequence of rectangular patterns of black and white tiles. Patterns at all stages have a height of 3 tiles and a black tile "core."

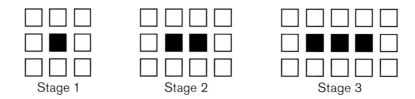

| Stage 1 | Stage 2 | Stage 3 |

a. Stage 1 has a core consisting of 1 black tile with a surrounding frame of 8 white tiles. Stage 2 has a core consisting of 2 black tiles with a surrounding frame of 10 white tiles.

b. The length of the pattern changes from one stage to the next. This means that the number of black tiles and the number of white tiles both change, since each new black tile adds two new white tiles—one above and one below it. The number of black tiles is always equal to the number of the stage, and thus, the number of black tiles increases by 1 from one stage to the next, while the number of white tiles increases by 2, adding one white tile above, and one white tile below, as described. The number of white tiles (W) at any stage is equal to two times the number of black tiles (B), plus 3 white tiles at the left end of the pattern and 3 whites tiles at the right end. In other words,

$$W = 2B + 6.$$

The total number of tiles in any pattern is equal to $W + B$, or $2B + 6 + B$, or $3B + 6$. At all stages, the pattern has a height of 3 tiles, as well as 3 white tiles on the left end, 3 white tiles on the right end, and a black tile core with paired white tiles above and below it.

c–d. Stage 11 has 11 black tiles, since the number of black tiles is equal to the number of the stage. It has $3(11) + 6$, or 39 tiles in all, since it has three rows of 11 tiles before adding in the 3 tiles on the left end and the 3 tiles on the right end, for a total of 39 tiles. Stage 11 thus has $39 - 11$, or 28, white tiles.

Solutions for "Take a Look at Those Squares"

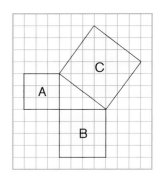

1. The right triangle with squares constructed on its sides is shown. Middle school students probably will not recognize the 3-4-5 right triangle. They can easily determine the areas of squares A and B, whose sides lie on grid lines. Determining the area of square C may be more difficult. However, because the activity sheet shows a centimeter grid, the students can measure the side of square C with a ruler with metric calibrations. Alternatively, they can determine the area of the square by decomposing it along grid lines into four congruent right triangles enclosing a unit square, as discussed in the text (see p. 115 and fig. 6.3).

   a. Area of square A = 9 square units

      Area of square B = 16 square units

      Area of square C = 25 square units

   b. Side length of square A = 3 units

      Side length of square B = 4 units

      Side length of square C = 5 units

   c. Students' responses will vary, but many students may notice that the sum of the areas of squares A and B is equal to the area of square C (or they may have heard of the Pythagorean theorem).

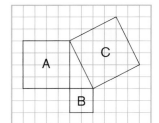

2. The right triangle with constructed squares is shown. Because the activity sheet shows a centimeter grid, students can measure the side of square C with a metric ruler, or they can determine the area by decomposing the square along grid lines into four congruent triangles and an enclosed square (see fig. 6.3).

   a. Area of square A = 16 square units

      Area of square B = 4 square units

      Area of square C = 20 square units

   b. Side length of square A = 4 units
      Side length of square B = 2 units
      Side length of square C = $\sqrt{20}$, or $2\sqrt{5}$, or approximately 4.472, units

   c. Students' responses will vary. As before, many students may notice that the sum of the areas of squares A and B is equal to the area of square C (or they may have heard of the Pythagorean theorem).

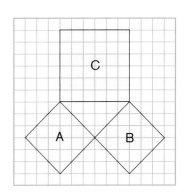

3. The right triangle with constructed squares is shown. The activity sheet shows a centimeter grid, so students can measure the sides of squares A and B with a metric ruler, or they can determine the areas of these congruent squares by decomposing each (or just one, for the sake of efficiency) along grid lines into four congruent triangles (see fig. 6.5).

   a. Area of square A = 18 square units

      Area of square B = 18 square units

      Area of square C = 36 square units

   b. Side length of square A = $\sqrt{18}$, or $3\sqrt{2}$, or approximately 4.243, units

      Side length of square B = $\sqrt{18}$, or $3\sqrt{2}$, or approximately 4.243, units

      Side length of square C = 6 units

   c. Students' responses will vary. As before, many students may notice that the sum of the areas of squares A and B is equal to the area of square C (or they may have heard of the Pythagorean theorem).

4. The right triangle with constructed squares is shown. The activity sheet shows a centimeter grid, so students can measure the sides of the squares with a metric ruler, or they can determine their areas by decomposing each square along grid lines into four congruent triangles enclosing a square (see fig. 6.5). The following calculations are based on determinations of the areas of the congruent triangles and the enclosed square in the case of each square (A, B, or C).

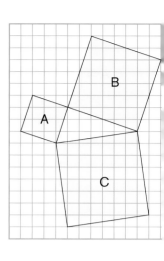

a. Area of square A $= \left(\frac{1 \times 3}{2} \times 4\right) + (2 \times 2) = 6 + 4 = 10$ square units

Area of square B $= \left(\frac{6 \times 2}{2} \times 4\right) + (4 \times 4) = 24 + 16 = 40$ square units

Area of square C $= \left(\frac{1 \times 7}{2} \times 4\right) + (6 \times 6) = 14 + 36 = 50$ square units

b. Side length of square A $= \sqrt{10}$, or approximately 3.162, units

Side length of square B $= \sqrt{40}$, or $2\sqrt{10}$, or approximately 6.325, units

Side length of square C $= \sqrt{50}$, or $5\sqrt{2}$, or approximately 7.071, units

c. Students' responses will vary. As before, many students may notice that the sum of the areas of squares A and B is equal to the area of square C (or they may have heard of the Pythagorean theorem).

5. *a–b.* Students' triangles will vary, as will the squares constructed on their sides. Figure 6.7 shows a table like that on the activity sheet, with sample values obtained from the applet Squaring the Triangle. Students should observe that the sum of the areas of the squares on the legs of each right triangle is equal to the area of the square on the hypotenuse.

6. Students' statements of the Pythagorean theorem will vary in expression but should convey the basic idea: The sum of the areas of squares constructed on the legs of a right triangle is equal to the area of a square constructed on the hypotenuse.

Solutions for "Can You Prove It?"

1–3. Students should cut out the four sets of congruent triangles carefully. When they arrange each set on grid paper to form a square whose side length is *a* + *b*, the figures that they produce and then draw will look like those on the right (also shown in fig. 6.9).

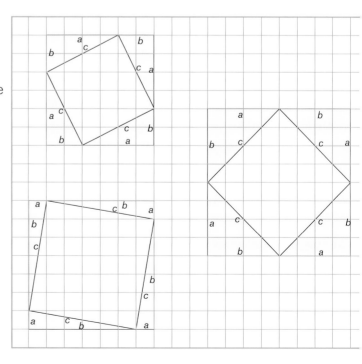

4–6. a. $A = (a+b) \times (a+b) = a^2 + 2ab + b^2$.

b. $A = \left(4 \times \dfrac{a \times b}{2}\right) + c \times c = 2ab + c^2$

c. $a^2 + 2ab + b^2 = 2ab + c^2$

$\quad\quad a^2 + b^2 = c^2$.

If your students are not sufficiently advanced in their study of algebra to write the algebraic expressions and perform the algebraic manipulations shown above, guide them in using their sets of triangles to decompose the area of a square that is $(a + b)$ units on a side in a different way, as shown at the right (art also appears as fig. 6.10).

This approach, discussed on pages 121–22, allows the students to use geometry to demonstrate that $a^2 + b^2 = c^2$ in the general case.

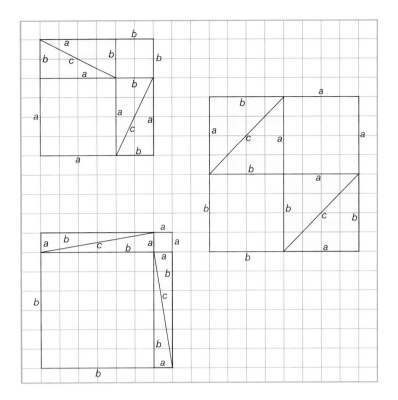

7. Students' responses will vary, but they should see that they now know with certainty that the Pythagorean theorem is true. Moreover, they should understand that the proof that they now have is much more valuable than the collection of examples that they had at the conclusion of "Take a Look at Those Squares." They should recognize that no matter how many examples they collect, the examples cannot serve to prove a theorem, although a single counterexample would disprove it.

References

Bakker, Arthur. "Reasoning about Shape as a Pattern in Variability." *Statistics Education Research Journal* 3 (November 2004): 64–83. http://www.stat.auckland.ac.nz/~iase/serj/SERJ3(2)_Bakker.pdf.

Ben-Zvi, David, and Joan Garfield. "Statistical Literacy, Reasoning, and Thinking: Goals, Definitions, and Challenges." In *The Challenge of Developing Statistical Literacy, Reasoning and Thinking*, edited by David Ben-Zvi and Joan Garfield, pp. 3–15. Boston, Mass.: Kluwer Academic Publishers, 2004.

Bishop, Joyce W., Albert D. Otto, and Cheryl A. Lubinski. "Promoting Algebraic Reasoning Using Students' Thinking." *Mathematics Teaching in the Middle School* 6 (May 2001): 508–14.

Blanton, Maria L., and James J. Kaput. "Instructional Contexts That Support Students' Transition from Arithmetic to Algebraic Reasoning: Elements of Tasks and Culture." In *Everyday Matters in Science and Mathematics: Studies of Complex Classroom Events*, edited by Ricardo Nemirovsky, Ann S. Rosebery, Jesse Solomon, and Beth Warren, pp. 211–34. Mahwah, N.J.: Lawrence Erlbaum, 2004.

Boats, Jeffery J., Nancy K. Dwyer, Sharon Laing, and Mark P. Fratella. "Geometric Conjectures: The Importance of Counterexamples." *Mathematics Teaching in the Middle School* 9 (December 2003): 210–15.

Bright, George W., Wallece Brewer, Kay McClain, and Edward S. Mooney. *Navigating through Data Analysis in Grades 6–8. Principles and Standards for School Mathematics* Navigations Series. Reston, Va.: National Council of Teachers of Mathematics, 2003.

Bright, George W., Patricia Lamphere Jordan, Carol Malloy, and Tad Watanabe. *Navigating through Measurement in Grades 6–8. Principles and Standards for School Mathematics* Navigations Series. Reston, Va.: National Council of Teachers of Mathematics, 2005.

Burton, Grace M. *Number Sense and Operations. Curriculum and Evaluation Standards for School Mathematics* Addenda Series, Grades K–6. Reston, Va.: National Council of Teachers of Mathematics, 1993.

Cai, Jinfa, and Stephen Hwang. "Generalized and Generative Thinking in US and Chinese Students' Mathematical Problem Solving and Problem Posing." *Journal of Mathematical Behavior* 21 (November 2002): 401–21.

Carroll, William M. "Middle School Students' Reasoning about Geometric Situations." *Mathematics Teaching in the Middle School* 3 (March/April 1998): 398–403.

Chapin, Suzanne H. "Mathematical Investigations—Powerful Learning Situations." *Mathematics Teaching in the Middle School* 3 (February 1998): 332–38.

Cuevas, Gilbert J., and Karol Yeatts. *Navigating through Algebra in Grades 3–5. Principles and Standards for School Mathematics* Navigations Series. Reston, Va.: National Council of Teachers of Mathematics, 2001.

Cuoco, Al, E. Paul Goldenberg, and June Mark. "Habits of Mind: An Organizing Principle for Mathematics Curriculum." *Journal of Mathematical Behavior* 15 (December 1996): 375–402.

Doyle, Walter. "Academic Work." *Review of Educational Research* 53 (Summer 1983): 159–99.

———. "Work in Mathematics Classes: The Context of Student Thinking during Instruction." *Educational Psychologist* 23 (March 1988): 167–80.

Driscoll, Mark. *Fostering Algebraic Thinking: A Guide for Teachers, Grades 6–10.* Education Development Center. Portsmouth, N.H.: Heinemann, 1999.

 Enderson, Mary C. "Using Measurement to Develop Mathematical Reasoning at the Middle and High School Levels." In *Learning and Teaching Measurement*, 2003 Yearbook of the National Council of Teachers of Mathematics (NCTM), edited by Douglas H. Clements, pp. 271–81. Reston, Va.: NCTM, 2003.

Famighetti, Robert. *World Almanac and Book of Facts 2000.* Edited by World Almanac. New York: St. Martin's Press, 2000.

 Friel, Susan N. "Teaching Statistics: What's Average?" In *The Teaching and Learning of Algorithms in School Mathematics*, 1998 Yearbook of the National Council of Teachers of Mathematics (NCTM), edited by Lorna J. Morrow, pp. 208–17. Reston, Va.: NCTM, 1998.

Friel, Susan, Sid Rachlin, and Dot Doyle, with Claire Nygard, David Pugalee, and Mark Ellis. *Navigating through Algebra in Grades 6–8. Principles and Standards for School Mathematics* Navigations Series. Reston, Va.: National Council of Teachers of Mathematics, 2001.

Goldin, Gerald A., and C. Edwin McClintock. "The Theme of Symmetry in Problem Solving." In *Problem Solving in School Mathematics*, 1980 Yearbook of the National Council of Teachers of Mathematics (NCTM), edited by Stephen Krulik, pp. 178–94. Reston, Va.: NCTM, 1980.

Greenes, Carole, Mary Cavanagh, Linda Dacey, Carol Findell, and Marian Small. *Navigating through Algebra in Prekindergarten–Grade 2. Principles and Standards for School Mathematics* Navigations Series. Reston, Va.: National Council of Teachers of Mathematics, 2001.

 Henningsen, Marjorie A. "Triumph through Adversity: Supporting High-Level Thinking." *Mathematics Teaching in the Middle School* 6 (December 2000): 244–48.

Kamii, Constance K. *Young Children Reinvent Arithmetic: Implications of Piaget's Theory.* New York: Teachers College Press, 1985.

 Lannin, John K. "Developing Algebraic Reasoning through Generalization." *Mathematics Teaching in the Middle School* 8 (March 2003): 342–48.

 Malloy, Carol. "The van Hiele Framework." On the CD-ROM for *Navigating through Geometry in Grades 6–8*, by David K. Pugalee, Jeffrey Frykholm, Art Johnson, Hannah Slovin, Carol Malloy, and Ron Preston. *Principles and Standards for School Mathematics* Navigations Series. Reston, Va.: National Council of Teachers of Mathematics, 2002.

McClain, Kay. "Reflecting on Students' Understanding of Data." *Mathematics Teaching in the Middle School* (March 1999): 374–80.

McClain, Kay, Paul Cobb, and Koeno Gravemeijer. "Supporting Students' Ways of Reasoning about Data." In *Learning Mathematics for a New Century*, 2000 Yearbook of the National Council of Teachers of Mathematics (NCTM), edited by Maurice J. Burke, pp. 174–187. Reston, Va.: NCTM, 2000.

Mooney, Edward S. "A Framework for Characterizing Middle School Students' Statistical Thinking." *Mathematical Thinking and Learning* 4, no. 1 (2002): 23–63.

Moore, David S. *Statistics: Concepts and Controversies.* 3rd ed. New York: W. H. Freeman, 1991.

National Center for Education Statistics. *Teaching Mathematics in Seven Countries: Results from the TIMSS Video Study.* Washington, D.C.: U.S. Department of Education, 2003.

National Council of Teachers of Mathematics (NCTM). *Curriculum and Evaluation Standard for School Mathematics.* Reston, Va.: NCTM, 1989.

———. *Professional Standards for Teaching Mathematics.* Reston, Va.: NCTM, 1991.

———. *Developing Mathematical Reasoning.* 1999 Yearbook of the National Council of Teachers of Mathematics, edited by Lee V. Stiff. Reston, Va.: NCTM, 1999.

———. *Principles and Standards for School Mathematics.* Reston, Va.: NCTM, 2000.

———. *Mathematics Teaching Today: Improving Practice, Improving Student Learning.* 2nd ed. Original title *Professional Standards for Teaching Mathematics.* Reston, Va.: NCTM, 2007.

Pugalee, David K., Jeff Frykholm, Art Johnson, Hannah Slovin, Carol Malloy, and Ron Preston. *Navigating through Geometry in Grades 6–8. Principles and Standards for School Mathematics* Navigations Series. Reston, Va,: National Council of Teachers of Mathematics, 2002.

Rachlin, Sid. "Learning to See the Wind." *Mathematics Teaching in the Middle School* 3 (May 1998): 470–73.

Raimi, Ralph. *A Mathematical Manifesto*, Part 2. New York: NYC HOLD, 2002, http://www.nychold.com/raimi-reason02.html.

Reys, Barbara J. Developing Number Sense. *Curriculum and Evaluation Standards for School Mathematics* Addenda Series, Grades 5–8. Reston, Va.: National Council of Teachers of Mathematics, 1991.

Rosenstein, Joseph G., ed. *New Jersey Mathematics Curriculum Framework.* Trenton, N.J.: New Jersey Mathematics Coalition and New Jersey State Department of Education, 1996. http://dimacs.rutgers.edu/nj_math_coalition/framework.html.

Schaeffer, Richard L., Ann E. Watkins, and James M. Landwehr. "What Every High-School Graduate Should Know about Statistics." In *Reflections on Statistics: Learning, Teaching, and Assessment in Grades K–12*, edited by Susanne P. Lajoie, pp. 3–31. Mahwah, N.J.: Lawrence Erlbaum, 1998.

Sconyers, James M. "Proof and the Middle School Mathematics Student." *Mathematics Teaching in the Middle School* 1 (November/December 1995): 516–18.

Simon, Martin A., Ron Tzur, Karen Heinz, and Margaret Kinzel. "Explicating a Mechanism for Conceptual Learning: Elaborating the Construct of Reflective Abstraction." *Journal for Research in Mathematics Education* 35 (November 2004), 305–29.

Smith, Margaret S., Mary Kay Stein, Fran Arbaugh, Catherine A. Brown, and Jennifer Mossgrove. "Characterizing the Cognitive Demands of Mathematics Tasks: A Task-Sorting Activity." In *Professional Development Guidebook for "Perspectives on the Teaching of Mathematics,"* companion booklet to the 2004 Yearbook of the National Council of Teachers of Mathematics (NCTM), edited by George W. Bright and Rheta N. Rubenstein, pp. 45–72. Reston, Va.: NCTM, 2004.

Smith, Margaret Schwan, and Mary Kay Stein. "Selecting and Creating Mathematical Tasks: From Research to Practice." *Mathematics Teaching in the Middle School* 3 (February 1998): 344–50.

 Steele, Diana F. "Using Schemas to Develop Algebraic Thinking." *Mathematics Teaching in the Middle School* 11 (August 2005): 40–46.

Stein, Mary Kay, Barbara W. Grover, and Marjorie Henningsen. "Building Student Capacity for Mathematical Thinking and Reasoning: An Analysis of Mathematical Tasks Used in Reform Classrooms." *American Educational Research Journal* 33 (Summer 1996): 455–88.

 Stein, Mary Kay, and Margaret Schwan Smith. "Mathematical Tasks as a Framework for Reflection: From Research to Practice." *Mathematics Teaching in the Middle School* 3 (January 1998): 268–75.

Stein, Mary Kay, Margaret S. Smith, Marjorie A. Henningsen, and Edward A. Smith. *Implementing Standards-Based Mathematics Instruction: A Casebook for Professional Development.* New York: Teachers College Press, 2000.

Stigler, James W., and James Hiebert. "Improving Mathematics Teaching." *Educational Leadership* 61 (February 2004): 12–16.

 Turner, Julianne C., Karen Rossman Styers, and Debra G. Daggs. "Encouraging Mathematical Thinking." *Mathematics Teaching in the Middle School* 3 (September 1997): 66–72.

Van de Walle, John. *Elementary and Middle School Mathematics: Teaching Developmentally.* 6th ed. Boston: Pearson Education, 2007.

Waring, Sue. *Can You Prove It? Developing Concepts of Proof in Primary and Secondary Schools.* Leicester, England: The Mathematical Association, 2000.